POWER SYSTEM PROTECTION

Volume 4: Digital protection and signalling

POWER SYSTEM PROTECTION

Volume 4: Digital protection and signalling

Edited by The Electricity Training Association

The Institution of Electrical Engineers

Published by: The Institution of Electrical Engineers, London,
United Kingdom

© 1995: Electricity Association Services Limited **EA**

Reprinted 1997

The Institution of Electrical Engineers,
Michael Faraday House,
Six Hills Way, Stevenage,
Herts. SG1 2AY, United Kingdom

British Library Cataloguing in Publication Data

A CIP catalogue record for this book
is available from the British Library

ISBN 0 85296 838 8

Printed in England by Short Run Press Ltd., Exeter

Contents

Foreword

The four volumes which make up this publication owe their origin to a correspondence tuition course launched in 1966 by the UK electricity supply industry, written by expert engineers from both the supply industry and manufacturers, and administered by the Electricity Council. The correspondence course continues to be provided to meet the needs of staff in the electricity supply industry throughout the world. Since privatisation of the industry in the UK the course is now provided by the Electricity Training Association, the industry's training organisation.

It became apparent soon after its inception that the work met a widespread need in the UK and overseas for a standard text on a specialised subject. Accordingly, the first edition of Power System Protection was published in book form in 1969 and has since come to be recognised as a comprehensive and valuable guide to concepts, practices and equipment in this important field of engineering. Because the books are designed not only to provide a grounding in the theory but to cover the range of applications, changes in protection technology mean that a process of updating is required. The second edition therefore presented a substantial revision of the original material and, although only minor changes have been made to the first three volumes, the publication of the fourth book in 1995, along with revised versions of existing works, reflects the considerable developments in the field of digital technology and protection systems.

The four revised volumes comprise 23 chapters, each with a bibliography. The aim remains that of providing sufficient knowledge of protection for those concerned with design, planning, construction and operation to understand the function of protection in those fields, and to meet the basic needs of an engineer intending to specialise in the subject.

In the use of symbols, abbreviations and diagram conventions, the aim has been to comply with British Standards.

The Electricity Training Association wishes to acknowledge the work both of the original authors, and of those new contributors who undertook the work of revising the first three volumes for this new edition; they are referred to at the head of the appropriate chapter. The Association also acknowledges the valuable assistance of National Grid Company, East Midlands Electricity,

Norweb, and Yorkshire Electricity staff and the help of the following in permitting reproduction of illustrations and other relevant material:

Allen West Company Limited, ASEA Brown Boverie Company Limited, BICC Limited, Electrical Apparatus Company, ERA Technology Limited, GEC Switchgear Limited, GEC Transformers Limited, Price and Belsham Limited, Reyrolle Protection Limited.

Extracts from certain British Standards are reproduced by permission of the British Standards Institution from whom copies of the complete standards may be obtained.

A particular indebtedness is acknowledged to the Chairman and members of the Editorial Panel, who directed and co-ordinated the work of revision for publication.

Chapter authors

R.K. Aggarwal	B Eng, PhD, C Eng, MIEE, SMIEEE
N. Ashton	C Eng, FIEE, ARTCS
K.A.J. Coates	C Eng, MIEE
P.C. Colbrook	BSc Tech, AMCST, C Eng, MIEE
L. Csuros	Dipl Eng, C Eng, FIEE
D. Day	BSc, C Eng, MIEE
P.M. Dolby	C Eng, MIEE
L.C.W. Frerk	BSc (Eng), C Eng, MIEE, FI Nuc E
F.W. Hamilton	B Eng, C Eng, FIEE
J. Harris	C Eng, MIEE
J.W. Hodgkiss	MSc Tech, C Eng, MIEE
L. Jackson	BSc, PhD, C Eng, FIEE
G.S.H. Jarrett	Wh Sc, BSc (Eng), C Eng, FIEE
M. Kaufmann	C Eng, FIEE
E.J. Mellor	TD
K.G. Mewes	Dip EE, C Eng, MIEE
P.J. Moore	B Eng, ACGI, PhD, C Eng, MIEE
J.H. Naylor	BSc (Eng), C Eng, FIEE, AMCT, DIC
C. Öhlén	MSc Electrical Engineering
H.S. Petch	BSc (Eng), MIEE, M Amer IEE
J. Rushton	PhD, FMCST, C Eng, FIEE
E.C. Smith	AMCT, C Eng, FIEE
C. Turner	DSc, F Inst P
H.W. Turner	BSc, F Inst P
J.C. Whittaker	BSc (Eng), C Eng, FIEE

Editorial panel

Protection symbols used in circuit diagrams

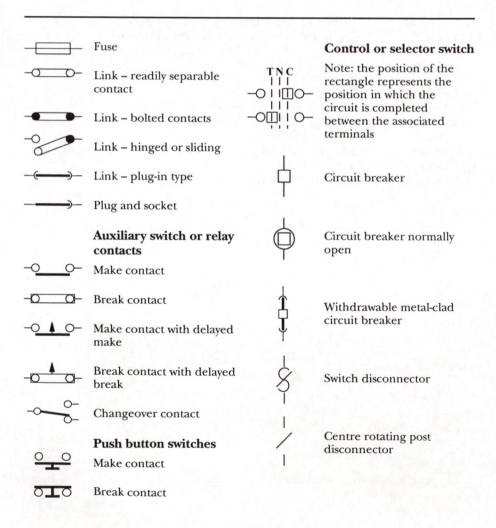

Fuse

Link – readily separable contact

Link – bolted contacts

Link – hinged or sliding

Link – plug-in type

Plug and socket

Auxiliary switch or relay contacts

Make contact

Break contact

Make contact with delayed make

Break contact with delayed break

Changeover contact

Push button switches

Make contact

Break contact

Control or selector switch

T N C

Note: the position of the rectangle represents the position in which the circuit is completed between the associated terminals

Circuit breaker

Circuit breaker normally open

Withdrawable metal-clad circuit breaker

Switch disconnector

Centre rotating post disconnector

Telephone type relay contacts

Make contact unit

Break contact unit

Changeover (break before make) contact unit

Changeover switch break before make

Make contactor

Mechanical coupling

Example – double pole contactor

Indirectly-heated bimetallic thermal element

Operating coil for contactors and relays – general

Coil with flag indicator

Series coil

Machine windings

General and shunt

Series

AC generator

Motor

Core (*if desired to indicate*)

Transformer or reactor winding

Single-break disconnector or earth switch

Additions to symbol for power-operated disconnector:
NA – Non-Automatic
A – Automatic

Fault throwing switch

Summation current transformer

Power and voltage transformers

Two winding

Simplified form

Auto-transformer

Simplified form – with delta tertiary

Transductor

Spark gap

Protective gas discharge tube

Capacitor

Power system protection

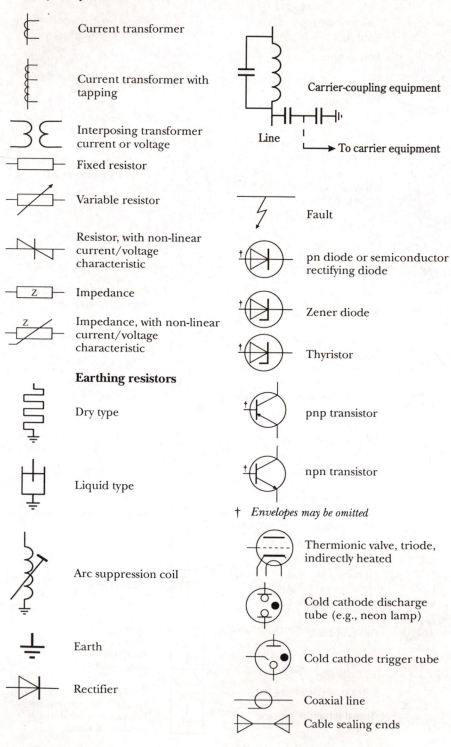

Current transformer

Current transformer with tapping

Interposing transformer current or voltage

Fixed resistor

Variable resistor

Resistor, with non-linear current/voltage characteristic

Impedance

Impedance, with non-linear current/voltage characteristic

Earthing resistors

Dry type

Liquid type

Arc suppression coil

Earth

Rectifier

Carrier-coupling equipment

Line

To carrier equipment

Fault

pn diode or semiconductor rectifying diode

Zener diode

Thyristor

pnp transistor

npn transistor

† *Envelopes may be omitted*

Thermionic valve, triode, indirectly heated

Cold cathode discharge tube (e.g., neon lamp)

Cold cathode trigger tube

Coaxial line

Cable sealing ends

Symbol	Description
	Rectifier equipment in bridge connection
	Amplifier
B	Buchholz – single float
B	Buchholz – two float
WT	Winding temperature single switch
WT	Winding temperature double switch
HSA	High speed ammeter
	Alarm flag relay
	Trip flag relay
R	Relay – general symbol
FP	Feeder protection
PLC PC	Power line carrier phase comparison protection
Z	Distance protection
MHO	High speed distance (Mho) protection
	Electric bell
	Signal lamp
E	Electromotive force (emf)
PP	Private pilot protection
POP	Post Office pilot protection
TP	Transformer protection
DB	Biased differential protection
PB D	Plain balance differential protection
T HVC	Transformer HV connection protection
CC	Circulating current
BB	Busbar protection
MCP	Mesh corner protection
HAR	High speed auto-close relay
OV	Overvoltage relay
X	High speed distance (reactance) protection (inverse definite minimum time)
3OC I	Three-pole overcurrent relay (inverse definite minimum time
2OC I EI	Two-pole overcurrent and single pole earth fault relayse (inverse definite minimum time)

OC I	One-pole overcurrent relay (inverse definite minimum time)
EI	Earth-fault relay (inverse definite minimum time)
3OC 2S I	Three-pole two stage overcurrent relay (inverse definite minimum time)
3OC DI	Three-pole directional overcurrent relay (inverse definite minimum time)
3OC	Three-pole overcurrent relay (instantaneous)
3OC XI	Three-pole overcurrent relay (extremely inverse definite minimum time)
E	Earth-fault relay (instantaneous)
3OC HS	Three-pole high set overcurrent relay
E SB LTI	Standby earth-fault relay (long time inverse definite minimum time)
E2S SB LTI	Two-stage standby earth-fault relay (long time inverse definite minimum time)
RP	Reverse power relay
E RES	Restricted earth-fault relay

T	Tripping relay
INT	Intertrip relay
INT S	Intertrip relay (send)
INT R	Intertrip relay (receive)
TD	Definite time relay
NEG PH SEQ	Negative phase sequence
LE	Lost excitation
3OC VC I	Three-pole voltage-controlled overcurrent relay (inverse definite minimum time)
3OC INT I	Three-pole overcurrent interlocked relay (inverse definite minimum time)
BF C CK	Breaker fail current check
LVC	LV connection protection

Chapter 1
Digital technology
Dr P. J. Moore

Introduction

This section is an introduction to the hardware components found in computer-based systems. Little prior digital knowledge is assumed. The chapter begins by investigating general digital circuits which leads to a relevant introduction to the subject of microprocessors, including peripheral devices, programming and analogue to digital conversion. The chapter ends with a brief look at specialist microprocessors.

1.1 Logic devices

All digital or logic circuits are based upon a branch of mathematics discovered by George Boole. The resulting theory has become known as *Boolean Algebra* in which only two numbers or states are allowed. These states are numerously referred to as high/low, on/off, 0/1 or true/false. The importance in digital circuit design is that Boolean functions may be realised by simple logic circuits. The most basic Boolean functions are:

$$\textbf{AND} - A \cdot B$$
$$\textbf{OR} - A + B$$
$$\textbf{XOR} - A \oplus B$$
$$\textbf{NOT} - \overline{A}$$

where, for the case of the AND function, the notation A.B represents the result from ANDing together the two Boolean variables A and B. The following expressions will be found useful:

$$A \cdot A = A \qquad A + A = A$$
$$A \cdot \overline{A} = 0 \qquad A + \overline{A} = 1$$
$$1 \cdot A = A \qquad A + 1 = 1$$
$$0 \cdot A = 0 \qquad A + 0 = A$$

Several simple logic circuits are depicted in Figure 1.1 which shows the symbol adopted for the logic circuit. In general these have two inputs A and B and one output X; the *truth table* is a written representation of how each logic circuit behaves according to its Boolean function. Note that new logic functions NAND and NOR have been introduced which are formed by NOTing the result of an AND or OR operation, respectively. It is common practice to refer to these logic circuits as *gates*.

AND

A	B	X
0	0	0
0	1	0
1	0	0
1	1	1

OR

A	B	X
0	0	0
0	1	1
1	0	1
1	1	1

NAND

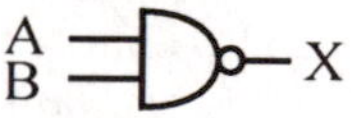

A	B	X
0	0	1
0	1	1
1	0	1
1	1	0

NOR

A	B	X
0	0	1
0	1	0
1	0	0
1	1	0

INV (Inverter)

A	X
0	1
1	0

XOR (exclusive OR)

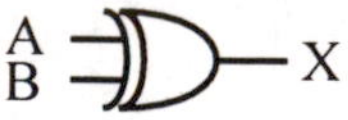

A	B	X
0	0	0
0	1	1
1	0	1
1	1	0

Figure 1.1   Simple logic gates

Logic gates may be connected together to form complex digital circuits. There are two distinct types of logic gate circuits, *combinational* circuits and *sequential* circuits. In combinational circuits the output of the circuit is uniquely defined by its inputs; this type of circuit has no memory action. A circuit having a memory action will give an output which depends upon the *sequence* in which the inputs are applied, thus these circuits are referred to as sequential circuits.

A simple combinational circuit is shown in Figure 1.2 and is seen to consist of simple logic gates. By inspection the logic function expression for this circuit can be written as:

$$F = \bar{B}.C + \bar{A}.B.C + A.B.C$$

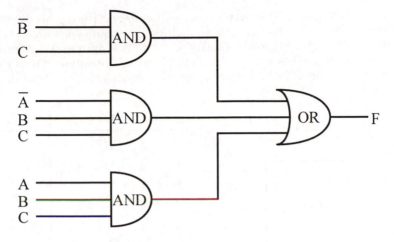

Figure 1.2   Simple combinational circuit

More importantly the converse is true; given a logic expression, it is possible to design a logic gate realisation. It is not immediately obvious from Figure 1.2 whether or not the realised circuit is efficient, i.e. uses the minimum number of logic gates. However, Boolean algebra allows this:

$$F = (A + \bar{A}) . B . C + \bar{B} . C$$
$$F = B . C + \bar{B} . C$$
$$F = (B + \bar{B}) . C$$
$$F = C$$

showing that there is, indeed, a much simpler realisation!

Another method of Boolean reduction is to use a pictorial representation called a *Karnaugh map*:

		A.B			
		00	01	11	10
C	0	0	0	0	0
	1	1	1	1	1

Karnaugh maps show all possible logic states for a Boolean expression pictorially. It should be clear from examination of the above map that the Boolean expression reduces to *F=C*. Similarly, less obvious Boolean reductions are possible by visual inspection. Karnaugh maps are useful for combinational circuits where there are normally 4 inputs. For larger numbers of inputs, designs are made by computer through the use of *computer aided design* (CAD) packages.

The memory action in a sequential circuit is provided by a logic device called a *flip-flop*. Several types of flip-flop exist but the most widely used variety is the

Data or D type flip-flop. The circuit representation and truth table are shown in Figure 1.3. Note that the truth table includes an input which changes state with time – this input is usually referred to as a *clock* input and will, in some form, always be found in sequential circuits. Note that the output of the flip-flop becomes a function of its data input from the previous clock cycle; it behaves as a *memory*.

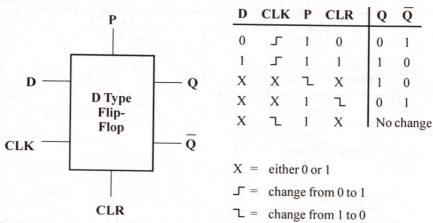

D	CLK	P	CLR	Q	Q̄
0	⌐	1	0	0	1
1	⌐	1	1	1	0
X	X	⌐	X	1	0
X	X	1	⌐	0	1
X	⌐	1	X	No change	

X = either 0 or 1

⌐ = change from 0 to 1

⌐ = change from 1 to 0

Figure 1.3   D type flip-flop

A simple sequential circuit is shown in Figure 1.4 and is seen to consist of both simple logic gates and flip-flops. Sequential circuits cannot be represented by simple Boolean algebra or Karnaugh maps. Instead a *state transition diagram* is used to show the possible states of the circuit. A state transition diagram for the sequential circuit of Figure 1.4 is shown in Figure 1.5.

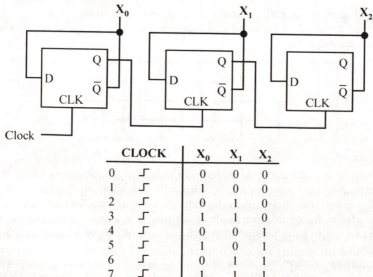

CLOCK		X_0	X_1	X_2
0	⌐	0	0	0
1	⌐	1	0	0
2	⌐	0	1	0
3	⌐	1	1	0
4	⌐	0	0	1
5	⌐	1	0	1
6	⌐	0	1	1
7	⌐	1	1	1

Figure 1.4   Simple sequential circuit

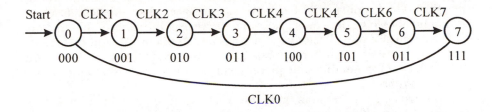

Figure 1.5   State transition diagram

In the early days of logic circuit design, each gate was made from discrete components, usually resistors, diodes and transistors. Progressively, logic gates became available in integrated circuit form where all of the constituent components were integrated onto the same piece of silicon. Such integrated circuits used conventional transistors and the logic gates became known as *transistor-transistor logic* or TTL. TTL was developed during the 1960s.

Another technology for integrated circuits was developed during the 1970s based upon *metal oxide semiconductors* (MOS). Although being sensitive to static discharge MOS is easily fabricated into integrated circuits. When used in *complementary* form (complementary metal oxide semiconductor, or CMOS), MOS can realise logic gates. CMOS enjoys an advantage of low power consumption over TTL technology, although TTL logic gates are usually quicker to operate than CMOS types.

The first NAND gates were produced in the early 1960s and these integrated circuits consisted of 30 to 40 transistors fabricated on silicon. Today integrated circuits contain upwards of 1 million transistors integrated onto a single piece of silicon – such integration is referred to as *very large scale integration* or VLSI. It may now be appreciated how, from the humble origins of simple AND and OR gates and flip-flops, sophisticated microprocessors may be derived. It is the colossal growth in silicon circuit fabrication technology over the last 30 years that has lead to the evolution of numeric protection relays amongst many other uses of digital circuitry.

1.2 Microprocessors

1.2.1 Historical development

Complex logic systems may be built up from arrangements of combinational and sequential logic circuits; very complex systems will lead to large and complicated circuits. Such systems are referred to as being *hardwired*, that is to say the underlying logic tasks performed by the system are embedded in the hardware of the system. This hardwiring could be in the form of a printed circuit board or, in the case of a fully integrated logic system, will be at silicon level. In either case, if a change has to be made to the logic structure, it will be very difficult to effect this change. Thus, as a means of increasing the utility of logic devices, it became desirable to produce logic systems that could be *programmed*, i.e. allowing the logic performed by the device to be arbitrarily specified, yet keeping the hardware unchanged. The highest level of programmable logic devices are known as *computers*. Initially computers were realised from discrete collections of logic circuits. As silicon integration became more and more dense, it became possible to integrate all the required logic devices for a computer onto one piece of silicon; these devices became known as *microprocessors*.

Although microprocessors may be realised from combinational and sequential forms of the previously described logic circuits, it is not necessary to know explicitly how this realisation is performed but rather to understand each building block of a microprocessor in its own right.

It should be noted that, since a microprocessor is a programmable device, it is incapable of performing any useful task until a suitable program, or sequence of instructions, has been provided for it. An *instruction* is a specification, made by the user, to make the microprocessor perform a certain logic function. Typically, an instruction is performed at every clock cycle. Complex tasks are built up from sequences of differing instructions. Although instructions may be logical in nature, i.e. behaving like the Boolean functions described in Section 1.1, they may also be arithmetic, i.e. adding, subtracting, dividing or multiplying numbers together, or simply instructions to exchange data to or from the outside world.

Figure 1.6 (opposite) shows a block diagram of the essential constituents of a microprocessor. A microprocessor has its own internal memory locations, referred to as *registers*, but, usually, it is also connected to additional external memory.

1.2.2 Basic operation

In order to communicate with memory, or any other device in the outside world, the microprocessor has distinct collections of signals which, when grouped together, are referred to as *buses*. Data to or from the outside world is carried on a *data* bus. Similarly, to access a specific part of the outside world an *address* bus is used. Several other signals are used in controlling these transfers and will be referred to as a *control* bus. The two common types of devices that would be immediately connected to a microprocessor are memory and input-output (IO) devices. The main IO devices found in numeric protection relays are described in Section 1.3.

There are two distinct types of memory that are used with microprocessors: *program* memory, which is used to store the sequence of instructions to be executed by the microprocessor, and *data* memory, which is used to store any pieces of information that are required in the execution of the task of the microprocessor. To clarify this point, an example of a word processing program run on a desk-top computer will be considered. In this case the program memory will contain the sequence of instructions which allow the computer to behave as a word processor, and the data memory will contain the document which the word processor is editing.

The basic operation of a microprocessor may be described as follows. The *program counter* contains the position in the program memory relating to the

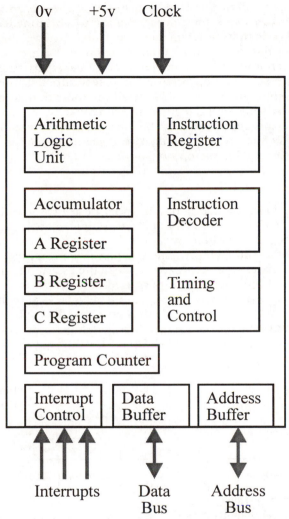

0v   +5v   Clock

Figure 1.6   Basic outline of a microprocessor

current instruction being executed. After completion of an instruction, the program counter is incremented and the timing and control circuitry arrange for the next instruction to be accessed from program memory via the address bus and fetched into the *instruction register* via the data bus. The contents of the instruction register are analysed by the *instruction decoder* which, in turn, instructs the timing and control circuitry to perform the desired task. Instructions can vary from simple read or writes of memory, in which case the address and data buses are used, to mathematical or logical manipulations of the internal registers, in which case the arithmetic and logic unit is used. Note that all blocks within the microprocessor are connected by internal buses. If it is necessary for an event in the outside world to be brought to the attention of the microprocessor, then this may be achieved via the *interrupt* inputs. An interrupt input can cause the microprocessor to go immediately to some specific part of code in its program memory.

The range of numbers that an internal register of a microprocessor can represent is determined by the number of individual *binary digits*, or bits, that the registers contain. Microprocessors have typically 8, 16 or even 32 bit number representation. It is possible for all microprocessors to handle very large numbers by using several registers to represent one number, e.g. an 8 bit microprocessor can represent a 32 bit number by the use of 4 registers. However, a 32 bit microprocessor will allow easier and faster processing of 32 bit numbers than an 8 bit microprocessor.

1.2.3 Memory devices

A very simplified representation of a memory device is shown in Figure 1.7. This device is seen to be composed of 8 rows of information each of which is capable of storing 4 lines of binary signals. In computer terms this is referred to as an 8 location 4 bit memory. The data bus of the computer is connected to the data lines of the memory. To select one of the 8 locations of the memory, the address bus of the microprocessor is used. However, it is seen that only 3 address lines are used to reference 8 locations. This is because the number of permutations of 3 binary digits is $2^3 = 8$. The actual decoding of the address lines to an individual memory location is handled within the memory chip itself (using a combinational logic circuit called a 3 line to 8 way decoder).

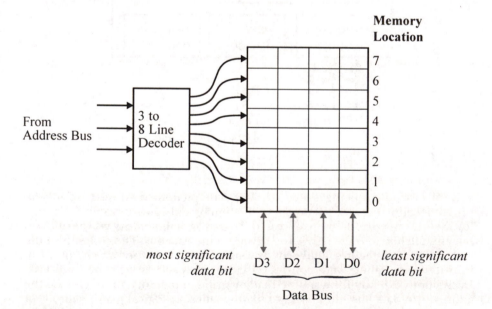

Figure 1.7   Simple memory device

So far no mention has been made of the control bus; in the context of a memory transfer, one signal within the control bus (not shown in Figure 1.7) is used to inform the memory chip whether the data is going to or from the microprocessor. If data is going *to* the microprocessor, then a memory *read* is occuring, similarly, data going to the memory *from* the microprocessor is referred to as a memory *write*.

In general, program and data memory are implemented with two different types of integrated circuit. Firstly, *read only memory* (ROM) is used for program memory applications. Its name implies that the microprocessor may only read the data stored within the ROM, thus making it useful for program memory uses. The data is entered into the ROM either by the manufacturer when the chip is fabricated or during a once-only electrical programming.

ROMs will retain their data even if the computer is switched off. Data memory is implemented from *random access memory* (RAM) – a term dating back to the early days of computers. In essence a RAM will allow the microprocessor to read its data, similar to a ROM, but will also allow the microprocessor to store new data in any of its locations. Thus RAMs are used as a transient store for data. RAMs are sometimes, and more correctly, referred to as read/write memories. In general RAMs lose their data contents if the microprocessor is powered down. A glossary of terms applying to ROMs and RAMs is given here:

- *ROM* – A factory-programmed ROM has its data introduced by the manufacturer during the silicon fabrication process. Generally only found in high volume production runs.
- *PROM* – Programmable ROM. ROM contents are programmed electrically by a once-only process involving special PROM programming equipment. Not necessarily performed by manufacturer. PROM programming is always much slower than RAM programming.
- *EPROM* – Electrically programmable ROM. This may be electrically programmed many times since ROM contents are erased by exposure to ultra-violet light. A quartz window is provided on the chip to facilitate this. Programming is the same as the PROM method.
- *SRAM* – Static RAM. The storage of 1 bit is implemented by a flip-flop. If undisturbed, data contents are retained for as long as power is applied to chip.
- *DRAM* – Dynamic RAM. The storage of 1 bit is effected using a capacitor. Data must be refreshed, typically every 1ms, due to charge leakage in the silicon substrate. This leads to extra circuitry and access to the data is slower than for SRAM. However, DRAMs are usually cheaper and are available with larger memory contents per chip than SRAMs since the fabrication of a capacitor uses less area on the silicon wafer than a flip-flop.
- *NVRAM* – Non-volatile RAM – i.e. a RAM which does not lose its contents after being powered down. Typically implemented as an SRAM with lithium battery included in the package. Data retention is guaranteed for as long as battery lasts, typically 10 years. For this reason, NVRAMs are unlikely to be found in numeric relays.
- E^2PROM – Electrically erasable PROM. Unlike an EPROM, this device does not need exposure to ultra-violet light in order to erase memory contents. Individual locations within the memory may be electrically erased and reprogrammed with the chip *in situ*. Can be used for storing relay settings.

- *Flash memory* – Similar idea, but different technology, to E^{2}PROM, i.e. memory may be electrically erased and programmed without resiting chip. Flash memory is available in larger memory size per chip than E^{2}PROM making it an attractive option to relay manufacturers for program memory store. Erasure of flash memory is restricted to large fractions of the on-chip memory rather than individual locations. Flash memory is likely to feature highly in future numeric relays for storing program memory and relay setting data. The flexibility of this device simplifies the shop-floor production of numeric relays, as well as allowing for easy on-site upgrades of relay software.

1.2.4 Binary number representation

Computers store numbers in binary notation. An example 8 bit number is 10110010 where the leading '1' is referred to as the *most significant bit*, or msb, and the trailing '0' is the *least significant bit* or lsb. To convert a binary number into decimal, it is necessary to calculate the decimal equivalent of each binary position. A simple way of encoding unsigned binary numbers is given below:

msb	2^7	2^6	2^5	2^4	2^3	2^2	2^1	2^0	lsb
	=	=	=	=	=	=	=	=	
	128	64	32	16	8	4	2	1	
	1	0	1	1	0	0	1	0	

	0
	2
	0
	0
	16
	32
	0
	128
TOTAL	**178**

And so 10110010 binary is seen to be 178 in decimal. Using this representation, the highest number is 255 decimal. Since it is laborious to refer continually to binary numbers explicitly, a shorthand notation has been developed which splits 8 bit binary numbers into two 4 bit 'nibbles' (note also that 8 bits are usually referred to as 'bytes' and 2 bytes, on a 16 bit microprocessor, are called a 'word').

Since each nibble can represent one of 2^4 = 16 different values, it is referenced by a single character where the first ten values are represented by 0–9 and the next six are represented by A, B, C, D, E and F. This system is called *hexadecimal* and 10110010 binary is B2 in hexadecimal. Note that a hexadecimal representation gives no indication of what the binary number really represents, it merely makes handling binary numbers easier for humans.

To allow negative numbers, the *twos complement* representation is used. In this representation, the msb is taken as negative:

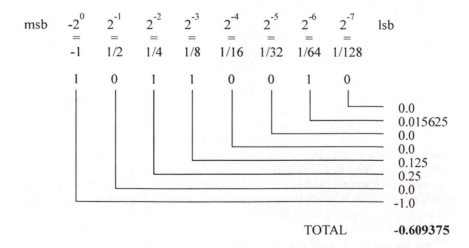

msb	-2^7	2^6	2^5	2^4	2^3	2^2	2^1	2^0	lsb
	=	=	=	=	=	=	=	=	
	-128	64	32	16	8	4	2	1	
	1	0	1	1	0	0	1	0	

	0
	2
	0
	0
	16
	32
	0
	-128

TOTAL   **-78**

Twos complement allows numbers in the range -128 to +127. Strictly speaking this is 'integer' twos complement. Another form is *fractional* twos complement:

msb	-2^0	2^{-1}	2^{-2}	2^{-3}	2^{-4}	2^{-5}	2^{-6}	2^{-7}	lsb
	=	=	=	=	=	=	=	=	
	-1	1/2	1/4	1/8	1/16	1/32	1/64	1/128	
	1	0	1	1	0	0	1	0	

	0.0
	0.015625
	0.0
	0.0
	0.125
	0.25
	0.0
	-1.0

TOTAL   **-0.609375**

An 8 bit fractional twos complement number can represent between -1 and 0.9921875. The advantage of fractional over integer arithmetic is seen if the numbers shown above are squared. For the integer case $(-78)^2 = 6084$, which is outside the range of the 8 bit representation. However, any fractional number can be squared to yield a representable result.

Fractional arithmetic is *fixed point* since the decimal point never moves; all numbers are less than or equal to 1. Finally, an increase in dynamic range may be achieved with the use of *floating point* representation. Here, numbers are represented in the form: $M \times 10^E$. Typically the *mantissa*, M, is represented with 10 bits and the *exponent*, E, is represented with 6 bits thus using 16 bits in total. The dynamic range is hence $\pm 512 \times 10^{\pm 32}$. The penalty of using floating point arithmetic is the extra processing required to perform simple operations such as addition, multiplication etc. compared to fixed point arithmetic. At the time of writing, no numeric relays use floating point arithmetic. However, this reflects upon the current high cost of microprocessors which can process floating point numbers at the speeds required for protection. It is likely that future relays will use this number representation.

1.2.5 Programming

To perform its intended task, a microprocessor needs to be programmed. Microprocessors may be programmed at three basic levels:

- *Machine code programming.* Here the specific codes to provide a given instruction, e.g. reading the contents of a RAM location, are individually evaluated by the programmer. This method of programming is very time-consuming and tedious and is mentioned here for completeness. Suitable only for very simple programs.
- *Assembly language programming.* This allows the machine codes to be addressed symbolically by a simple mnemonic. The programmer writes a series of mnemonics to perform the desired task. This program is then *assembled* into the relevant machine codes using another computer program called an *assembler*. This method is quicker to develop than machine code programming but is still tedious for long programs. However, resulting code is very efficient. Parts of numeric relay programs, where it is essential that the microprocessor executes the code as quickly as possible, are programmed in assembly language.
- *High level language programming.* A computer program written in a *high level language* such as Basic, Fortran, C, Pascal etc is far easier to understand than an assembly language program since the programs read similar to English. It is thus easier to develop the code and far simpler to write long programs. However, the resulting code is generally less efficient than for the previous two methods, i.e. the computer will take a longer time to execute a given task. Statements written in the high level language are converted into machine code by a program called a *compiler*. The non-time critical parts of a relay program are usually programmed in a high level language.

1.3 Analogue to digital conversion

1.3.1 Introduction

To gain information on the state of the power system, a numeric protection relay takes regular samples from the secondary voltage and/or current signals applied to it. This process is referred to as *analogue to digital conversion* and is perform by special hardware. In practice, useful power system signals are bipolar, i.e. either positive or negative, and are converted into digital form using twos complement representation. To simplify the following discussion on conversion, analogue signals are taken to be unipolar, i.e. positive. However, the general principles may be extended to include bipolar conversion.

1.3.2 Digital to analogue converters

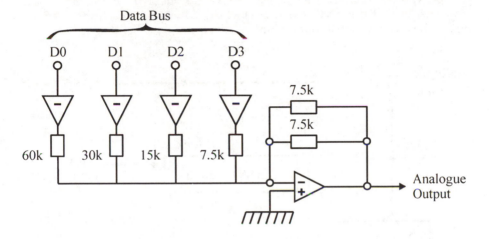

Figure 1.8   4 bit digital to analogue converter

To aid the understanding of analogue to digital conversion, firstly a *digital to analogue converter* (DAC) will be described. A basic circuit for a 4 bit DAC is shown in Figure 1.8 which is seen to consist of an operational amplifier connected to buffered data line inputs via weighting resistors. It is seen that the weighting resistors connected to each of the data lines increase in binary progression. The gain of the operational amplifier is given by:

$$\frac{V_{out}}{V_{in}} = -\frac{R_f}{R_w}$$

where V_{in} is the output voltage of the buffering amplifiers connected to the data lines (note this will only be one of two possible voltages), V_{out} is the analogue output voltage, R_f is the feedback resistor across the operational amplifier (in this case the parallel combination of two 7.5k ohm resistors) and R_w is the value of the weighting resistor if a data line is active. The negative gain of the amplifier is compensated by making the buffering amplifiers inverting. If more than one data line is active, then the analogue output voltage is the sum of the two V_{out}s calculated from the above expression. Thus the voltage at the output is directly proportional to the binary number represented by the data lines. The feedback resistors across the operational amplifier ensure that the largest binary number corresponds to highest analogue voltage. By increasing the number of data line inputs, and progressively increasing the series resistor values, DACs of 8, 10, 12, 14 and 16 bits may be implemented.

1.3.3 Analogue to digital converters: ramp converters

A simple *analogue to digital converter* (ADC), the so-called *ramp converter*, is shown in Figure 1.9 and consists of a comparator, binary counter, an AND gate, clock input and a DAC. A comparator is similar to an operational amplifier and, as used in this circuit, will give an output of '1', or high, if the analogue input voltage is greater than the DAC output, and an output of '0', or low, otherwise. Note, because of the AND gate, the clock signal will not reach the counter unless the comparator output is high.

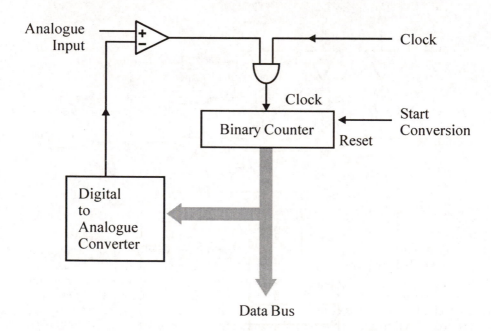

Figure 1.9   Ramp converter

With the desired analogue voltage present on the analogue input, the *start conversion* signal is given which resets the binary counter. Thus the DAC output is zero, the comparator output is high and, in turn, the clock pulses will pass into the counter. Assuming the analogue input to be non-zero, as the counter increments, the output of the DAC appears as a ramp. When the DAC voltage exceeds the analogue input voltage, the comparator output will go low and prevent the clock pulses reaching the counter. Thus the binary counter digital output lines now hold the nearest digital value corresponding to the analogue input.

The ramp ADC is rarely used in practice since the conversion time, the time required to ascertain the digital equivalent value of the input analogue signal, increases with the input voltage. To make the conversion fast, very high speed clock pulses are required which introduce other problems.

1.3.4 Analogue to digital converters: successive approximation converters

An adaptation to the ramp converter which produces a far superior ADC is the *successive approximation converter*, shown in Figure 1.10. The binary counter is replaced by a slightly more complicated logic arrangement which will be referred to as a successive approximation register. Figure 1.11 (overleaf) shows the output of the internal DAC and the analogue input signal as a 4 bit successive approximation converter makes a conversion. Prior to the conversion, all data lines are set to zero. It is seen that on the first clock cycle the DAC goes to half of its maximum output, this is equivalent to setting high the most significant bit of the data bus, D3. On the next clock cycle the successive approximation logic senses that the DAC output is still lower than the input signal since the comparator output is high. Thus, the D2 data line is now set high.

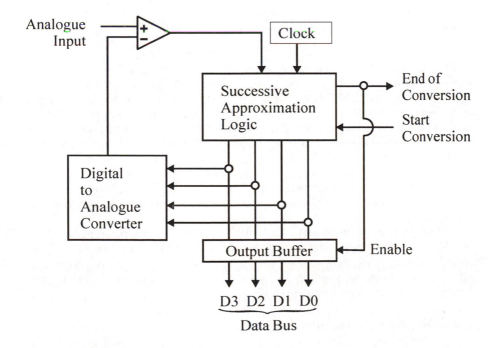

Figure 1.10 Successive approximation converter

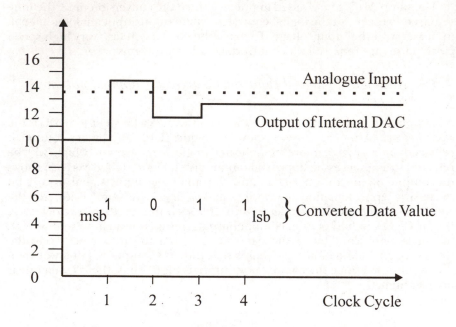

Figure 1.11 Operation of successive approximation converter

This results in the comparator going low, implying that the digital equivalent is greater than the analogue input, so, on the third clock cycle, D2 is set low again and D1 is set high. The fourth clock cycle results in D0 being set high and yields the digital result: 1011. It is seen that a successive approximation ADC individually tests each bit of the output data lines in turn; thus the time taken to convert is always fixed at the number of bits multiplied by the internal clock period. Conversion times for successive approximation ADCs are in the region of 15–30µs. ADCs usually have an 'end-of-conversion' signal which is commonly connected to an interrupt line of the microprocessor to inform when a conversion has finished and that the converted value is available for processing.

Although other types of ADC are available, successive approximation ADCs are the preferred type for protection relays. Other types include flash converters which are designed to have very fast conversion times (<1µs) and are consequenctly very expensive. Such conversion times are currently considered unnecessary in a protection relay.

1.3.5 Sample and hold amplifiers

Successive approximation ADCs are commonly found in digital relays. Since the successive approximation ADC takes, say, 25µs to convert the input signal, it is likely that some change in the input signal will occur during the conversion time. To eliminate this source of error, the input analogue signals are passed through *sample and hold* (SH) amplifiers which, upon command from the microprocessor, hold the input signal at a constant analogue level for the duration of the conversion. This is shown in Figure 1.12 where, during normal operation – the *sample* mode – the switch is closed and the output of the SH amplifier follows the input. Immediately prior to a conversion being made, the microprocessor opens the switch and the previous analogue voltage is held on the capacitor and thus the output remains constant – the *hold* mode. The two amplifiers in Figure 1.12 are present to buffer the SH circuit from the effects of the rest of the analogue input stages.

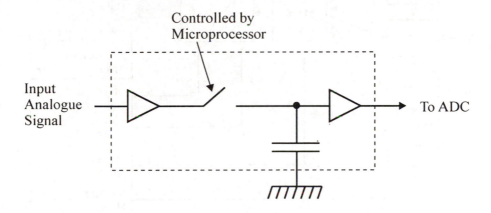

Figure 1.12 Sample and hold amplifier

1.3.6 Multiplexers

Successive approximation ADCs are still relatively expensive and it is uncommon to find more than one such ADC used in a relay (note that continuing development in large scale integration techniques may well render successive approximation ADCs much cheaper in the future and correspondingly influence numeric relay hardware). In relays requiring multi-channel inputs, such as a distance relay where a minimum of 6 channels (3 voltage, 3 current) are used, a device called a *multiplexer* is used to switch each of the input channels to the input of the ADC sequentially. However, for a 25μs conversion time ADC, it will be seen that the sixth input channel is converted 150μs after the first channel has begun. At 50Hz, a 150μs delay is equivalent to a phase shift of 2.7° and is thus a potential source of error to the relay algorithm. Thus, it is common for each analogue input channel to have its own SH amplifier. A typical digital relay analogue input stage is shown in Figure 1.13. Note that the multiplexer (MUX) is under the control of the microprocessor and that the SH amplifiers are all selected simultaneously into either the sample or hold modes.

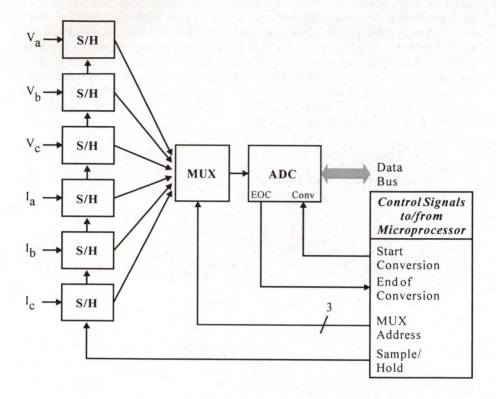

Figure 1.13 Relay analogue to digital conversion arrangement

1.3.7 Analogue to digital conversion in protection relays

Successive approximation ADCs are available in 8, 10, 12, 14 and 16 bit types. The number of bits in the ADC influences the dynamic range of the signal which the ADC converts. For example, an 8 bit ADC can digitise an analogue signal into 1 of $2^8 = 256$ different digital levels. The correct number of bits for a given relay depends upon the application. A relay will work correctly with an over-specified ADC, i.e. greater number of bits than strictly necessary, but not the other way around. An example is given here of how an ADC may be specified for a distance relay.

Consider a distance relay which has a minimum setting impedance of 4 Ω (relative to the relay). For a distance relay, the current rather than the voltage inputs will have the greater dynamic range. The minimum setting impedance will correspond to the highest current level which is, assuming a secondary VT of 110V, 110/4 = 27.5A. However, it is possible that the current may be offset during a fault, therefore, allowing for 100% offset, the maximum expected current is 55A. Suppose that the relay must operate for a minimum current level of 25mA and this can be represented by 1 digital level. Hence, the dynamic range for one polarity of the current is $55/0.025 = 2200$. Hence for a bipolar signal, the dynamic range is 4400. The ADC which comes closest to this figure is a 12 bit type which allows for a dynamic range of $2^{12} = 4096$. However, to ensure that the ADC covers the specification, a 14 bit converter must be used which has a dynamic range of $2^{14} = 16\,384$. Thus, a 14 bit converter would be required for this situation. In general, most high performance numeric relays use 12, 14 or 16 bit ADCs.

1.4 Specialised microprocessors

Typically, numeric relays use 12 or 14 bit ADCs. Prior to the introduction of 16 bit microprocessors, 8 bit microprocessors were standard, but not ideally suited to processing 12 or 14 bit values. Thus, the advent of 16 bit microprocessors was instrumental in the progress of numeric relays. One of the first 16 bit processors on the market was the Intel 8086 which was launched in 1978. The 8086 series is one of the most successful family of processors ever produced due to its widespread use in IBM PCs (usually as the 80286, 80386 or 80486). Another popular processor launched around the same time as the 8086 was the Motorola 68000 which, like the 8086, has been continually updated.

Although the emerging processor technology was suited to the development of numeric relays, there still existed a problem of digital multiplication. Any numeric relay, other than the most simple application, will execute a large number of multiplications whilst performing its protection function. Since multiplication on a microprocessor was, at the time, achieved by a series of shifts and additions which took a relatively large number of clock cycles to execute, this led to relay algorithms being very conservative in the number of multiplications used since multiplications used up processing time. Note that all the calculations performed by the relay must be completed in the time between ADC samples. A typical multiplication time for a standard 16 bit microprocessor in 1982 was $10\mu s$.

It was common at this time to see prospective relay designs that used simple digital filter coefficients, such as $1/2$ and $1/4$, (see Section 2.3 'Digital filtering') which can be achieved with single instruction shift operations. A stop-gap in the problem of digital multiplication was the emergence of hardware multipliers (HMs), single VLSI chips which were dedicated to the task of multiplying 16 bit numbers and could produce a result in, typically, 100ns. HMs were originally launched in the mid 1970s but it took several years before prices dropped significantly. A drawback to using HMs was their high power consumption (5W per chip) but this was solved by the launch of CMOS HMs in 1982. Despite the fast multiplication time of the HM, the effective multiplication rate was a lot slower since the microprocessor had to spend time sending data to and from the HM.

A significant step forward was made by the launch, in 1983, of the Texas Instruments TMS320 range of 16 bit digital signal processors. The TMS320 differs from conventional 16 bit microprocessors by having a hardware multiplier integrated directly onto its chip. The architecture of the TMS320 range was especially designed for digital signal processing (DSP), for example it is possible to perform a multiplication and an addition, a common operation in DSP as will be shown later, in one instruction cycle. Early versions of the TMS320 had an instruction cycle of 200ns but this rapidly decreased to 100ns. Over the past 10 years, there has been a steady growth in the performance of digital signal processors and they are now available from several manufacturers. The following table summarises the state-of-the-art for 1993:

Table 1.1   *Digital signalling chip specifications*

Manufacturer	Texas Instruments	Analog Devices	Motorola	NEC
Part Number	TMS320C40	ADSP21060	DSP96002RC40	μPD77230
Word Length	32 bit	32 bit	32 bit	32 bit
Instruction cycle period	20 ns	25 ns	25 ns	150 ns
Floating point operation ?	✔	✔	✔	✔

Modern high-performance numeric relays are almost exclusively based upon digital signal processing chips.

1.5 Reference

1   HAZNEDAR, H.: 'Digital microelectronics' (Benjamin/Cummins, 1991)

Chapter 2
Digital signal processing
Dr P. J. Moore

Introduction

The processing of signals which have been converted to digital form – *digital signal processing* – is now commonplace and this subject area is likely to grow more important in the future. The purpose of this chapter is to give an elementary grounding in digital signal processing including the process and limitations of sampling, digital filtering and spectral analysis. Quite unusually for this subject, little reference is made to mathematics, and so there are limitations to the depths of explanation. Further details may be found in the references at the end of the chapter.

2.1 Continuous versus discrete waveforms

Inherent to digital signal processing is the representation of waveforms as a series of numbers. However, when this representation is used, the very nature of the waveform has been changed and it is important that this distinction is understood.

If a 50Hz waveform is displayed on a standard oscilloscope, then the trace of the waveform is said to be *continuous*; that is, at every point in time, there is a distinct value which represents the 50Hz waveform. If it is possible to examine a small section of the waveform in great detail, then it would always appear to be continuous, no matter how closely the waveform is examined. All power system waveforms are continuous, as are the waveforms from a microphone or a record player. Note that, in the context of electrical engineering, *analogue* waveforms are continuous.

When waveforms are converted to digital form, then they will no longer be continuous. Table 2.1 shows the digital representation, taken at intervals of 1ms, of the first half-cycle of a 50Hz waveform of peak amplitude 10V, where a digital value of 100 represents 1V. At, say, 3ms, the digital representation of the waveform is 809 and remains so until a time of 4ms at which point it changes to 951. The digital representation is said to be *discrete*, since over a period of, in this case, 1ms, there is only one discrete value which represents the original waveform. Another term for this representation is a *discrete time signal*. This is distinct from the continuous waveform representation of the sine wave where, between 3ms and 4ms, there are an infinite number of values, since a continuous waveform may be infinitely subdivided. Despite the apparent differences between the two waveform representations, the discrete values are said to be a *unique* representation of the original sine wave. That is to say that only the original sine wave, and not any other waveform, may be converted to produce the set of digital values shown in table 2.1. However, this may not always be the case, as will be shown in the next section.

Table 2.1   Digital representation of a sine wave

Time (ms)	0	1	2	3	4	5	6	7	8	9
Voltage	0	3.090	5.878	8.090	9.511	1.000	9.511	8.090	5.878	3.090
Digital value	0	309	588	809	951	1000	951	809	588	309

2.2 Sampling

The process by which continuous waveforms may be represented as discrete values is referred to as *sampling* and is performed by an analogue to digital converter, and sample and hold amplifier, as described in the previous chapter. In Table 2.1, the discrete values are valid for a period of 1ms, at which point a new sample is made – this implies that the *sampling frequency* is 1kHz. The reciprocal of sampling frequency is referred to as the sampling interval. The sampling frequency is not chosen arbitrarily and, in general, is a governing factor in the design of the digital protection relay hardware and will be discussed later. However, an important relationship exists between the sampling frequency and the frequency of the waveform to be sampled; this relationship is referred to as the *sampling theorem* (see, for example, References 1 or 2).

Succinctly stated, the sampling theorem says that the sampling frequency must be greater than twice the highest frequency to be sampled. If this rule is disobeyed, then the unique digital representation of the original continuous waveforms is lost and an effect called *aliasing* occurs. The effect of aliasing is that two different continuous waveforms, when sampled, can appear as the same digital representation. Although this may appear unlikely, a simple exercise will show this to be true. Figure 2.1 shows two continuous sine waveforms; one at a high frequency (a), and one at a low frequency (b). The vertical dotted lines represent the exact instant of sampling and, where a sampling operation has occurred on each of the sine waveforms, a square box is drawn to indicate that a sample has been made. If the sampling frequency is denoted as f_s, then waveform (a) has a relative frequency of $10/9 f_s$ which is, of course, contrary to the sampling theorem. Examination of the 'samples' of waveform (a) will reveal that the original waveform is lost; the 'samples' appear the same as a sine wave of frequency $1/9 f_s$. In fact the samples on waveform (a) are completely indistinguishable from samples taken from a waveform of $1/9f_s$. It is common parlance to say that (at a sampling frequency of f_s) the frequency of $10/9f_s$ has been *folded down*, or *aliased*, to a frequency of $1/9f_s$. The waveform in Figure 2.1(b) has a relative frequency of $1/3f_s$ which is in agreement with the sampling theorem and, it is clear to see, that the sampled sine wave is a good representation of its original continuous waveform. It does not matter that the samples of (b) do not closely approximate a sine wave; this is merely how a discrete time signal appears when drawn on a graph.

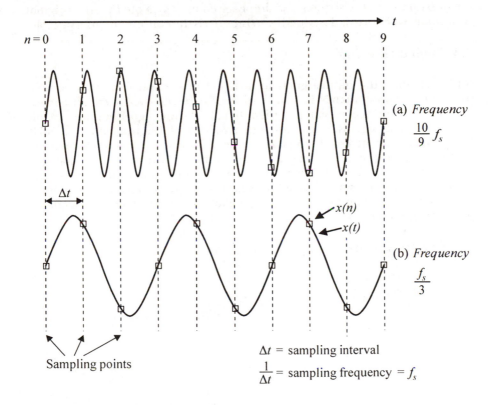

Figure 2.1   Effect of aliasing in sampled signals

In a digital protection relay, signals coming from capacitor voltage transformers and current transformers can contain, in addition to the 50Hz component, frequencies of tens of kHz under conditions of power system faults and switching operations. Since the sampling frequency is fixed by the relay hardware, it is necessary to ensure that the sampling theorem is obeyed by band-limiting all continuous signals entering the relay prior to sampling. This is achieved by the use of an analogue filter which is designed to remove any frequencies existing on the input signal which are greater than half the sampling frequency; such filters are referred to as *anti-aliasing filters*. Note that almost without exception, digital protection relays process only the power system frequency (i.e. 50Hz) information contained in the input signals. Thus, it is imperative that the anti-aliasing filter removes any frequencies that would 'fold down' to 50Hz after sampling.

When referring to discrete time signals, it is common to refer to a specific set of values as being a *sequence*. In Figure 2.1(b), the sequence of values of the sampled waveform x is described by $x(nT)$ where T is the sampling interval and n is an index which allows any value in the sequence to be specified. Commonly, T is omitted in this description since it is always implied and, as in Figure 2.1(b), the description of the sequence is simply given as $x(n)$. Note that the original continuous waveform is described as $x(t)$ where t is the continuous time variable.

2.3 Digital filtering

After sampling, the input signals to a digital protection relay may contain frequencies at other than 50Hz which will need to be removed to ensure correct relay operation since, as stated earlier, digital relaying algorithms are usually based on 50Hz signals. The most efficient way of filtering out non-50Hz components is by the use of a digital filter, i.e. a filtering process which operates on the sampled power system waveforms. Although the use of an analogue anti-aliasing filter was discussed earlier, it is far better to use digital filters rather than analogue types since digital filters have a shorter *group delay* which leads to shorter relay operating times. Group delay is the time taken for a signal to pass through the filter.

2.3.1 *Time domains and frequency domains*

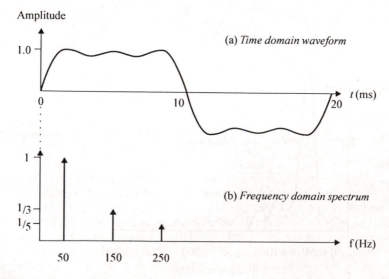

Figure 2.2   Time and frequency domains

Before discussing digital filters, it is important to draw a distinction between two types of description which are used in the context of filters and signals.

Figure 2.2 (a) shows a waveform which is clearly not a sine wave, in actual fact it is the result of superimposing the following sine waves:

(i) 50 Hz sine wave, magnitude 1
(ii) 50 Hz sine wave, magnitude 1/3
(iii) 250 Hz sine wave, magnitude 1/5

Figure 2.2(b) shows the above information expressed as a graph. Both Figures 2.2(a) and 2.2(b) describe essentially the same item. If it were necessary to describe the pertinent information contained in Figure 2.2 to someone else, then Figure 2.2(b), which is a *frequency domain* description is far easier to work with than Figure 2.2(a) which is a *time domain* description. Figure 2.3 contains another example of time and frequency domain information. Figure 2.3(a), which could represent a faulted power system voltage waveform, is another time

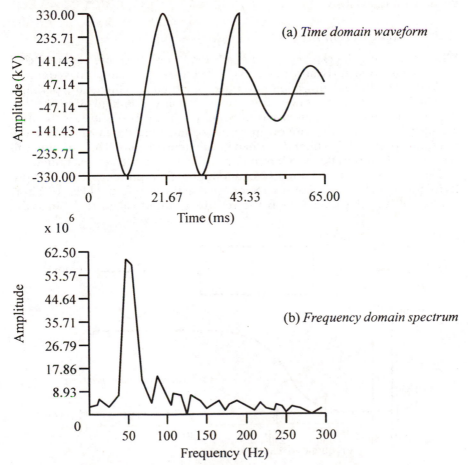

Figure 2.3   Time and frequency domains

domain description and its corresponding frequency domain description is shown in Figure 2.3(b). Clearly, it is preferable to describe the information, in this case, as a time domain description. The use of time domain and frequency domain representations is interchangeable and, in general, it is purely a matter of convenience that dictates which representation is used.

It can be seen from Figure 2.2 that knowledge of the frequency domain is sufficient to allow construction of the corresponding time domain. In fact, this is not entirely true in the general sense, since in the frequency domain, phase as well as magnitude information is required for a complete description. In Figure 2.2(b) all the components of the waveform are assumed to share the same value at $t=0$. Hence, if magnitude and phase information are available, then the time domain may be derived. The same is true in reverse, that frequency domain information may be derived from only a time domain description, although this is less easy to see intuitively. Techniques for transferring between time and frequency domains will be described in Section 2.4 on spectral analysis.

2.3.2 Filter descriptions

When describing the characteristics of a filter, it soon becomes apparent that the frequency domain is more or less exclusively used; for example it is common to describe a filter as being a low pass filter, or a high pass filter, and, perhaps, it would be simple to draw a frequency response graph to illustrate the description. However, as discussed in Section 2.3.1, it is possible, with equal accuracy, to make a description in terms of the time domain. And so the question arises, how is a filter described in the time domain? This question will be answered shortly after another important feature of filters is examined.

Figure 2.4 shows a filter and its frequency domain representation or, simply, its frequency magnitude response. Suppose that the waveform of Figure 2.2(a) is injected into this filter, how is the resulting output calculated? Since the

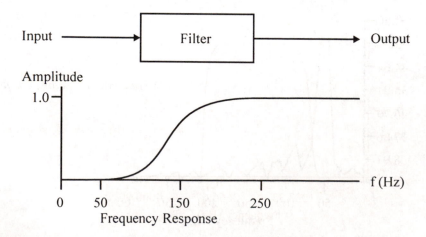

Figure 2.4   Filter and frequency response

frequency domain representation of the injected waveform is known (Figure 2.2(b)) it is possible to evaluate individually the response of each sine wave component of the waveform and then reconstruct the resulting time domain waveform. This process, which is impractical for most purposes, uses the frequency domain to calculate the time domain answer. There is a mathematical process which can be used to calculate the resulting filter output without any reference to the frequency domain. This process is referred to as *convolution* and it uses a characteristic of the filter called its *impulse response* in order to calculate the resulting output waveform. Thus, the output signal from the filter is result of *convolving* the input signal with the filter impulse response.

The impulse response is the time domain description of a filter which was queried earlier. The impulse response of a filter is as unique as its frequency response but, as Section 2.3.1 highlighted, there are situations where time domain descriptions are more useful than frequency domain descriptions. When describing filters, frequency responses are far more useful than impulse responses, as Figure 2.5 shows.

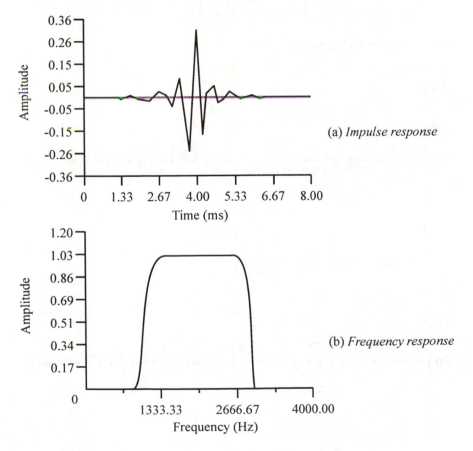

(a) *Impulse response*

(b) *Frequency response*

Figure 2.5   Frequency and impulse response of a filter

When dealing with analogue filters, it is difficult to understand the concepts of convolution or impulse responses by examination of the internal workings of an analogue filter which may consist of operational amplifiers, transistors, capacitors etc. Indeed, analogue convolution may only be described in terms of integral calculus (and will not be explained here) and the filter response to an impulse, a very narrow isolated pulse of great magnitude, can only be approximated. This in some way explains why the frequency domain is preferred for describing filters. However, when the internal mechanisms of digital filters are examined, the concepts of impulse responses and convolution are immediately apparent, hence the foregoing discussion. Furthermore, digital filters have very clearly defined impulse responses and the digital equivalent of convolution is easy to understand without having to resort to complicated mathematics.

(a) *Digital filter with applied unit impulse*

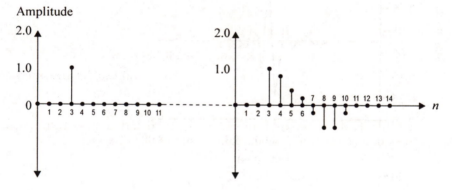

(b) *Same filter with larger impulse applied*

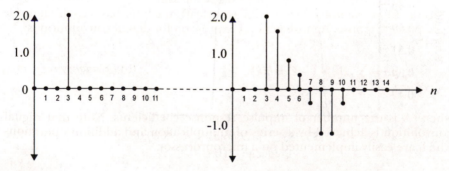

Figure 2.6   Frequency and impulse response of a filter

Figure 2.6(a) shows a digital filter with an input sequence consisting of many zeros with an isolated value of one in the middle. This particular input sequence is the digital equivalent of the impulse described earlier. Note that the impulse is as narrow as possible. The output of the filter, the impulse response, is also shown and is seen to consist, for this specific filter, of a sequence of 8 values which somewhat resemble a decaying sine wave. Figure 2.6(b) shows the same situation except that the impulse has been scaled up by a magnitude of 2. Note that the filter output is also scaled by a factor of 2, yet its other characteristics, such as shape and length of response, remain unchanged. In general, the output of a digital filter is the summation of the individual responses of the filter to each sample in the input sequence. Figure 2.7 shows this in more detail where a digital filter, with specified impulse response, has a sine wave based sequence of samples as its input. Each isolated response to the first 5 inputs samples is shown individually at the correct point in time; the filter output is simply the sum of all the relevant parts of the responses at a given point in time. Let the sequence of values of the impulse response be denoted by $h[k]$ for $k = $ 1,2,3,4 and 5. Note that the values of $h[k]$ are usually referred to as the *filter coefficients*. Also let the input sequence values be denoted by $x[n]$ for $n = 0,1,2,$ 3, ... , and the filter output sequence be denoted by $y[n]$ for $n = 0,1,2, 3, ...$ At sample $n=5$, the output of the filter can be seen to be:

$$y[5] = x[5] \times h[1] + x[4] \times h[2] + x[3] \times h[3]$$
$$+ x[2] \times h[4] + x[1] \times h[5] \tag{2.1}$$

In general, the output of the filter at sample n is given by:

$$y[n] = x[n] \times h[1] + x[n-1] \times h[2] + x[n-2] \times h[3]$$
$$+ x[n-3] \times h[4] + x[n-4] \times h[5] \tag{2.2}$$

This is essentially an equation for digital convolution and may be written in a more convenient form using the mathematical summation sign:

$$y[n] = \sum_{k=1}^{5} x[n+1-k] \times h[k] \tag{2.3}$$

Note that Equations 2.2 and 2.3 are exactly the same.

In the filter example of Figure 2.7 (overleaf), the filter has 5 coefficients in its impulse response. A more general expression for digital convolution is:

$$y[n] = \sum_{k=1}^{N} x[n+1-k] \times h[k] \tag{2.4}$$

where N is the number of impulse response coefficients. Note that digital convolution is achieved by a series of multiplication and addition operations which are easily implemented on a microprocessor.

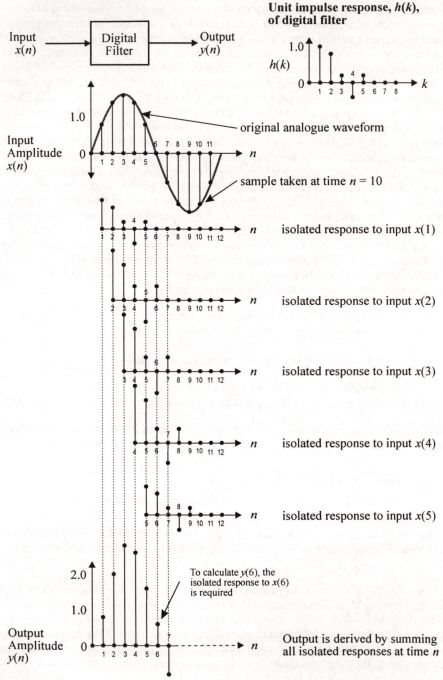

Figure 2.7   Graphical explanation of digital convolution

2.3.3 Types of digital filter

2.3.3.1 Finite impulse response

Equation 2.4 is understood to represent a digital filtering function where the input sequence of values to be filtered, $x[n]$, is convolved with the filter impulse response, $h[k]$, which consists of N coefficients. N is also referred to as the *filter length*. Although N can be large, it cannot be infinitely large since, in order to implement the filter, a total of N multiplications and N additions must be performed between sampling instants. A filter of this type is, thus, referred to as a *finite impulse response* (FIR) filter. FIR filters have the property that their group delays are never greater than NT where T is the sampling interval. Furthermore, when designing FIR filters, the group delay may be used as a design parameter. This is very important for digital protection relay applications since the group delay will directly influence the relay operating time and, generally, is kept to a minimum.

Although digital filters have been explained in relation to their impulse responses, when designing a digital filter, the starting point is a frequency response of the desired filter. The difficulty then is to find the impulse response which corresponds to the desired frequency response (the derivation of the frequency response from the impulse response is covered in Section 2.4.2, fast Fourier transform). Latterly, the design of FIR filters has been revolutionised by the development of an optimal filter design program by Parks and McLellan (see Reference 2 for more details), referred to as the *Remez exchange method*. The Remez exchange program is given the desired frequency response and the filter length from which the optimal filter coefficients are calculated. Ironically, this technique is of limited use in protection applications where filter lengths are of critical importance (see Section 2.5) and more empirical methods of filter design are usually employed by relay manufacturers.

2.3.3.2 Infinite impulse response

Another important class of digital filters have an infinite impulse response (IIR). IIR filters cannot be implemented via Equation 2.4. Instead an equation which uses both past values of the filter input and output is used:

$$y[n] = \sum_{k=1}^{M} x[n+1-k] \times a[k] + \sum_{k=1}^{M} y[n+1-k] \times b[k] \qquad (2.5)$$

where a and b are a set of M filter coefficients. Since previous values of the filter output are used in Equation 2.5, this equation is described as being *recursive*. The main drawback to the use of IIR filters in digital protection relays is that the group delay cannot be specified in the design process. This makes their use in protection somewhat onerous and, in general, FIR filters are usually the preferred type; however, there is an important exception to this which will be discussed in Section 2.5.1.

2.4 Spectral analysis

2.4.1 Discrete Fourier transform

Waveforms may be described in the time domain or, with equal accuracy, in the frequency domain. This leads to the question: is there a generalised method of deriving, say, the frequency domain description of a given time domain waveform? The answer to this question is yes, the technique to be used is called the Fourier transform, which is named after its inventor, Jean Baptiste Joseph Baron de Fourier. The process of moving between the time domain and the frequency domain, or vice-versa, is referred to as *transforming*. Quite often, the impetus for performing Fourier transforms is the need to have a knowledge of the frequency *spectrum* of some time domain waveform, hence the section title, spectral analysis. Figure 2.8 shows a selection of common time domain waveforms and their associated frequency domain representations. Note that the frequency domain graphs are shown as having both positive and negative frequency spectrums. When the Fourier transform is used for spectral analysis of time domain waveforms such as power system voltage and current waveforms, the positive and negative frequency information from the Fourier transform will be identical. Thus, it is easier to consider all frequency spectrums as being purely positive. The Fourier transform not only allows conversion from time to frequency domain but, also, through the use of the inverse Fourier transform, can convert from the frequency domain to the time domain. However, it is the former process of time to frequency conversion which is of interest in this context and further references to Fourier transforms will assume time to frequency conversion.

In common with the discussion on continuous convolution, the abstract use of Fourier transforms on continuous waveforms involves integral calculus (which, again, will not be described here). However, the Fourier transform may be adapted for use on discrete time signals – as such it is referred to as the *discrete Fourier transform* or DFT.

A typical use for the DFT is to analyse the time domain waveform of Figure 2.2(a) and hence calculate the frequency spectrum shown in Figure 2.2(b) together with the associated phase information. This is of particular relevance to protection where it is clearly advantageous to estimate the 50Hz component of a power system waveform corrupted by noise.

In practice, the DFT, for transforming from time to frequency domains, is implemented as two equations:

$$\text{Re}[X(m)] = \sum_{k=0}^{N-1} x(n-k) \cos\left(\frac{2\pi mk}{N}\right)$$

$$\text{Im}[X(m)] = \sum_{k=0}^{N-1} x(n-k) \sin\left(\frac{2\pi mk}{N}\right)$$

$$(2.6)$$

where N is the number of samples in the discrete time sequence $x(n)$, m is called the harmonic index, (it specifies which frequency the DFT will evaluate) and

Frequency Domain

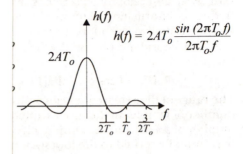

$$h(f) = 2AT_o \frac{\sin (2\pi T_o f)}{2\pi T_o f}$$

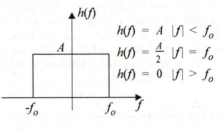

$$h(f) = A \quad |f| < f_o$$
$$h(f) = \frac{A}{2} \quad |f| = f_o$$
$$h(f) = 0 \quad |f| > f_o$$

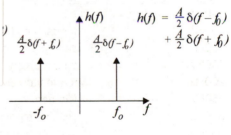

$$h(f) = \frac{A}{2}\delta(f-f_o)$$
$$+ \frac{A}{2}\delta(f+f_o)$$

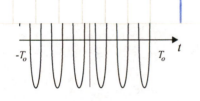

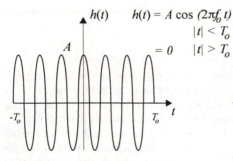

$$h(t) = A \cos (2\pi f_o t)$$
$$|t| < T_o$$
$$= 0 \quad |t| > T_o$$

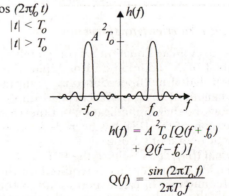

$$h(f) = A^2 T_o [Q(f+f_o)$$
$$+ Q(f-f_o)]$$

$$Q(f) = \frac{\sin (2\pi T_o f)}{2\pi T_o f}$$

Figure 2.8   Common Fourier transforms

$X(m)$ is the frequency component. Since sampling produces *discrete* time signals, so the DFT produces *discrete* frequencies. Note the similarity of Equation 2.6 with Equation 2.5 for discrete convolution. Equation 2.6 evaluates the frequency information in terms of real and imaginary components. These are simply related to the more familiar magnitude and phase description. For example, consider a phasor V, which has a magnitude of A and a phase of ϕ. Then:

$$\mathrm{Re}[V] = A \cos \phi \qquad \mathrm{Im}[V] = A \sin \phi$$

The range of discrete frequencies evaluated by the DFT is dependent on N, the number of samples in the input sequence (this is sometimes referred to as the number of *points*) and the sampling frequency f_s. All frequencies evaluated are harmonically related to the lowest, non d.c., frequency which is referred to as the *fundamental* frequency. The fundamental frequency is given by f_s/N. The *second harmonic* frequency is twice the fundamental frequency, the third harmonic is three times the fundamental etc. For example, consider an input sequence consisting of 20 samples taken at a sampling frequency of 1kHz. Thus, $N=20$, and $N/2=10$ discrete frequencies will be evaluated. Specifically these are shown in Table 2.2.

Table 2.2

Frequency	Description	Harmonic index
0Hz	d.c.	$m=0$
50Hz	fundamental	$m=1$
100Hz	second harmonic	$m=2$
150Hz	third harmonic	$m=3$
200Hz	fourth harmonic	$m=4$
250Hz	fifth harmonic	$m=5$
300Hz	sixth harmonic	$m=6$
350Hz	seventh harmonic	$m=7$
400Hz	eighth harmonic	$m=8$
450Hz	ninth harmonic	$m=9$

2.4.2 Fast Fourier transform

In order to evaluate Equation 2.6 completely for all discrete frequencies, there are a total of N^2 addition and multiplication operations to be performed. Although digital multiplication on a computer is now relatively quick, it is always beneficial to use efficient algorithms which reduce the number of operations, be they multiplication or otherwise. An efficient algorithm for the computation of the DFT has been developed which reduces the number of arithmetic operations required. This algorithm is referred to as the *fast Fourier transform* (FFT). The results of the FFT are identical to the results of the DFT, however, less computer time is required for the FFT. The FFT takes advantage of the redundancy of calculation found in the DFT. This redundancy is that the same operation, e.g. multiplying a sample by a coefficient, occurs several times when the entirety of the DFT is evaluated. The exact operation of the

FFT will not be discussed here, but a comprehensive description may be found in Reference 3. The number of arithmetic operations used by the FFT is $N \log_2 N$, with the restriction that N must be a power of 2. The FFT becomes more efficient as N increases. Table 2.3 reinforces this point:

Table 2.3

Number of samples	No. of operations DFT	No. of operations FFT
16	256	64
32	1024	160
64	4096	384
128	16 384	896
256	65 536	2048
512	262 144	4608

Although the FFT would appear to be far superior to the DFT in terms of computing performance, a factor which heavily influences digital protection relay design, it has one drawback which inhibits its use in protection. Simply, the FFT calculates **all** of the frequency spectrum; this may not always be required for the protection function. Indeed, as stated earlier, it is usually only the 50Hz component which is of interest, although occasionally other components may render useful information, e.g. the use of third harmonic restraint in generator stator earth fault protection. The DFT, however, may be evaluated for only one frequency component and hence it is the DFT equation, or its variant, which is found in protection applications.

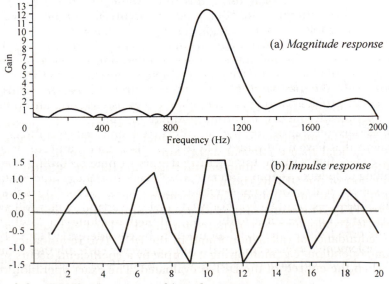

Figure 2.9   FIR filter frequency and impulse responses

In the context of digital protection relaying, the most useful application of the FFT is to evaluate the frequency response of a digital filter as a part of the design process. This is achieved by performing an FFT on the filter impulse response. An example of this is shown in Figure 2.9 where the frequency magnitude response of a 20 point impulse response has been evaluated by the use of a 512 point FFT. A large number of points for the FFT are used to give good frequency resolution, this is why the frequency response almost appears to be continuous. When the number of points in an impulse response is less than the number of points in the FFT, the unused points are set to zero.

2.5 Digital filtering in protection relays

2.5.1 Design constraints

The main impetus for using digital filtering in digital relays is, as stated earlier, that for a given frequency response, a digital filter will have a shorter group delay than an equivalent analogue filter which, in turn, leads to a faster relay operating time. In general, with FIR-type digital filters, there is no readily available analogue counterpart and thus the use of FIR digital filtering is an advantage in itself since frequency responses may be achieved that would otherwise be impossible.

However, the use of digital filtering in protection relays poses some problems. Consider the FIR filter impulse response of Figure 2.10(a) and its frequency response, Figure 2.10(b), that has been evaluated using a FFT assuming a sampling frequency of 4kHz. Since there are 80 points in the impulse response, the length of this filter, its group delay, would be $NT = N/f_s$ = 80/4000 = 20ms. Examination of the frequency response shows that it has a pass band centred on 50Hz and excellent rejection of harmonics. As such it would appear to be an ideal filter for a protection relay. However, due to its group delay of 20ms, a relay using this filter would be unable to operate correctly, and consistently, until 20ms after the occurrence of the fault and is consequently unable to give ultra-high-speed performance.

Figure 2.11 depicts a similar situation but with an impulse of 12 points. Thus, the group delay is 3ms and ultra-high-speed operation is now a possibility. However, examination of its frequency response reveals a less favourable situation with the passband of the filter centred at 260Hz. Thus the fundamental problem of digital relay filter design is choosing an optimum balance between frequency response and operating time.

The discussion regarding digital filters concentrated on filters with coefficients that have fixed values. There is another important class of filters called *adaptive* filters where the coefficients vary in time according to some controlling influence. Adaptive filters are useful because the characteristics of power system signal distortion under fault conditions vary according to the type of fault. Since the type of fault is not predictable, an adaptive filter will always give an optimal filter response to the fault, although in addition to filter group delay, the time for the filter to adapt needs to be considered. Adaptive filters can be implemented as either FIR or IIR types. However, currently there is much interest in the area of Kalman filtering, which is an IIR-based adaptive filter.

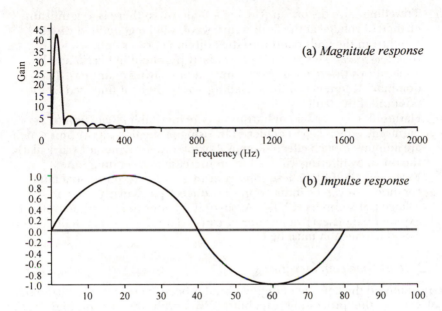

Figure 2.10 80 point sine wave based filter – frequency and impulse responses

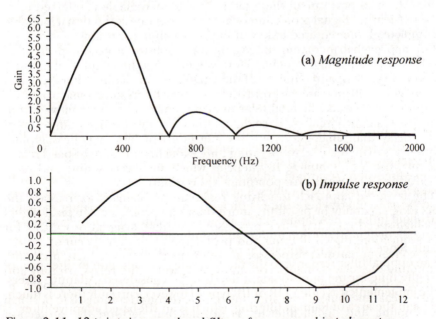

Figure 2.11 12 point sine wave based filter – frequency and impulse responses

The factors to be considered in designing digital filters for protection relays are:

(a)   Travelling wave distortion. During a fault, when there is a significant change of voltage at the fault point, waves will be generated which travel from the fault point into the adjoining power network – *travelling waves*. This effect is particularly prevalent in overhead ehv applications where it causes corruption of relaying quantities. The dominant frequency of the travelling waves will be influenced by the position of the fault.

(b)   Harmonics. Power system harmonics are inevitable. Rejection of harmonic effects is best achieved by using protection algorithms which are immune to such effects. Removal of low order harmonics, particularly the third, by filtering will lead to an increase in operating time.

(c)   Exponential offsets. These can occur in genuine response to a fault condition, or can be induced in transducers, particularly current and voltage transformers (CVTs). As such, it is better to remove actively all exponential offsets from whatever cause. This implies that digital filters with d.c. rejection must be used.

2.5.2   *Real-time considerations*

A constraint on the amount of digital signal processing performed by the relay is placed by the processing capability. Even with the use of high speed processors such as the TMS320 family as described in Section 1.4, problems can arise. Between each sample, the relay has to filter all the measured signals, run the protection algorithm and reach a decision regarding the fault. For example, with a sampling frequency of 4kHz, the relay has $1/4000 = 250\mu s$ within which to perform all these tasks. This time available is referred to as *real-time*. Clearly, a digital protection relay cannot operate correctly if it is unable to complete all the required tasks within the real-time constraint.

One approach to increasing the real-time available for protection processing is to lower the sampling frequency. However, lowering the sampling frequency also lowers the passband required of the analogue anti-aliasing filters and, as a consequence, will increase the group delay of these filters since a sharper cutoff is required. In essence the total relay filtering function has been moved from digital to analogue circuitry. Since digital filters are more efficient in terms of group delay than analogue types, the net effect of lowering the sampling frequency is to increase relay operating time. Note that ultra-high-speed relays, as a rule, use high sampling frequencies which allow very simple and basic anti-aliasing filters with correspondingly short group delays.

The favoured approach to solving real-time problems is to increase the processing capability by providing more than one processor. A typical digital ultra-high-speed relay has three processors, two digital signal processors for data acquisition, digital filtering and protection algorithm execution, and a more conventional microprocessor to perform the scheme logic including monitoring the relay elements, communicating with other relays, tripping and implementing time delays where appropriate. An alternative approach to increase real-time, but without affecting hardware considerations, is to perform the sampling and digital filtering as normal, but perform the protection algorithm at half the sampling frequency. This increases available real-time, but delays the operating time by only one sampling interval.

2.6 Further reading

1   LYNN, P. A.: 'An introduction to the analysis and processing of signals, 3rd Edition', (MacMillan,1989)

2   RABINER, L. R. and GOLD, B.: 'Theory and application of digital signal processing', (Prentice-Hall, 1975)

3   BRIGHAM, E. O.: 'The fast Fourier transform and its applications', (Prentice-Hall, 1988)

Chapter 3
Digital communications and fibre optics
C. Öhlén and Dr R. K. Aggarwal

Introduction

This chapter introduces the final element of background knowledge required for the remainder of this book. It begins with the general subject of digital communications, explaining transmission types, protocols, error handling and networks, and concludes with a description of the transmission medium of fibre optics, a technology which is increasingly found within the substation environment.

3.1 Digital data transmission

3.1.1 Introduction

Data transmission includes many different functions. It is therefore important first to answer the questions: *what* information, *when*, *where* and to *whom*, before saying *how*. In the field of substation automation we can divide the communication into information and commands. We can furthermore structure both the information and the commands into different levels of importance. This in turn will give the requirement for the accuracy or the quality of the information as well as *when* it should be sent. On-line information with short response time and high resolution is required for commands of breakers whereas statistical data is not time-critical. Instead of on-line information immediately transferred, the data can be regularly stored until requested.

To where the data (information and commands) are sent can basically be divided into four locations: internally on the bay level, centrally on the substation level, remotely to and from the control centre and remotely to and from the main office. Today, some utilities are implementing mobile workplaces for both the operator and engineer. The operator on duty can, for example, monitor the power system from home using a portable PC. For remote communication the available media is normally already given. This could be utility-owned microwave and power line carrier, or the telephone system. Today the available speed on these systems is normally between 300 and 2400 bps, although in some cases 9600 and 19 800 can be obtained. With the increased use of fibre optics and satellite communications the telephone companies are now able to offer data transmission at speeds greater than 1 Mbps.

The user of the information determines how it should be sent and presented. Too much information is often worse than too little. The information has to be processed and "user friendly". In most utilities the user can be divided into four personnel groups: Technician, Operator (i.e. Load dispatcher), Protection relay engineer and Service engineer, as depicted in Figure 3.1.

"Right information at the right time"

		Normal operation	Disturbance	Location
	Technician	Function test Temporary setting Check indications	Check alarms	Terminal
	Service engineer	Parameter settings Maintenance test Maintenance statistics	Replace modules	Terminal
	Operator	Functional control Alarm and report handling Load statistics	Restore operation	Control room
	Relay engineer	Parameter-calculations Verification (post-fault) Fault statistics	Analyse faults	Central office

Figure 3.1   Users of utility-provided information

Information to be transmitted is coded as a stream of bits, i.e. 0s or 1s. The most common types of codes are CCITT No. 5 (7 bits ASCII), CCITT No. 2 (Telex), CCITT S.61 (TELETEX), EBCDIC (IBM Extended Binary Coded Interchange Code). CCITT No. 5 and other types of ASCII are the most common type for data communication. A parity bit is normally added to the 7 bit characters which gives a total length of 8 bits. A 7 bit code allows a total of 128 characters to be represented; 96 of these are defined as alpha-numerical characters and the remaining 32 are used for different purposes such as control characters, e.g. carriage return. The ASCII character 'A', for example, is represented by the code 1000001, similarly, character '9' is 1001110.

3.1.2 Simplex, half duplex and full duplex transmission

The term *simplex* means that data is transferred in only one direction. This can be used, for example, in measuring applications where no control function is required. *Half duplex* means that data can be transferred in both directions, but only one way at the time. This is the most common type of data communication.

In many applications an interactive process is necessary which means no new data are sent until a response is received following earlier data. *Full duplex* means that data can be communicated in both directions at the same time. This method is used, for example, in the high-speed transfer of files between two computers. Half duplex is the method used in substation control systems (SCS).

3.1.3 Asynchronous and synchronous transmission

It is not sufficient for communication that the sending and receiving ends are using the same language. It is also necessary that both ends agree upon when a character starts and when it ends. This can either be done with asynchronous or synchronous transmission.

Asynchronous transmission means *not* synchronous, and so the time interval between two characters is undefined. Each character is started with a start bit and ended with a stop bit. Both the sending and the receiving end need individual synchronisation from their own clock. The receiving end detects when the start bit is received. Using its own clock frequency, which is approximately the same as the sending frequency, it will then read 8 bits. The 8th bit is a parity bit and comes before the stop bit. In asynchronous transmission, only a small amount of data, e.g. one character, can be sent between start and stop bits, otherwise differences in clock frequency between the receiving and sending ends will prevent the data being read.

During *synchronous* transmission the receiving end will copy the sending frequency using information contained within the incoming bits. This means that a large number of characters can be received without interruption of start and stop bits.

Asynchronous transmission is sent when it occurs. Synchronous transmission sends large data strings continuously. Both types are suited to different applications. Asynchronous transmission is common between a terminal and a hostcomputer, and synchronous transmission is common between two mastercomputers.

3.1.4 Error handling

Three basic types of error detection exist: parity checking (vertical redundancy), block check characters (horizontal redundancy) and CRC (cyclic redundancy checking).

In parity checking, each 7 bit character has a parity bit which is set to reflect whether there is an even or odd number of 1s in the character. 1000101 has an odd number of 1s and will therefore have the parity bit of 0, whilst 1010101, having an even number of 1s, will have the parity bit set to 1. As an alternative to this, block check characters, a whole block of 7 bit characters can be checked by examining the parity of all characters in a specified bit position, e.g. in a block of ten 7 bit characters, this check shows the parity of all most significant bits, etc. When the block is received and the parity of the block is correct, the sending end will receive an acknowledgement. If the parity is incorrect, then the block is resent. Finally CRC is typically a 16 bit checksum which is added in the end of each block. A 16 bit CRC will detect all errorbursts which are up to 16 bits or shorter and 99.998% of all longer errors.

3.1.5 Protocols and standards

A protocol is a set of rules regarding the exchange of information. Protocols are divided into several layers, each specifying different functions for both hardware and software. Open System Interconnect (OSI) is a proposed ISO model which was initially published in 1979, and today exists as ISO 7498 and CCITT X.200. The goal is to have a common terminology and set of rules for development of standards, thus standardising the interface between layers and allowing easier communications between equipments from differing manufacturers. The layers are 1: physical layer, 2: data link layer, 3: network layer, 4: transport layer, 5: session layer, 6: presentation layer and 7: application layer. Layers 1–4 deal with the transmission and layers 5–7 deal with the application.

Many protocols are therefore needed to specify communication. There are two basic types of protocols – one to specify the communication between two units within the same layer and one to specify the interface between two layers.

Today standards exist from various international bodies such as ISO, CCITT and ANSI. Industry associations such as IEEE, EIA and ECMA are also presenting recommendations and standards in addition to the *de facto* standards set by large manufacturers such as DEC and IBM. Despite this, at present, there is no specification for a common communication standard which freely allows interconnection between the equipment of different manufacturers. Furthermore, technological developments are moving so rapidly that new methods are being continuously introduced. It is however possible to communicate between different manufacturer's equipment if the two protocols are known. If the protocols are different, this can be achieved through the use of a *protocol convertor*, although this may be difficult depending on the different functionality and capacity of the protocols. Ideally the number of interconnections between the two equipments should be minimal, and in the same location, to keep the cost of protocol conversion low and the functionality high.

The *physical layer* has the objective to send data bits through the selected media. This involves communication both within a specific unit and between different units. Relevant questions to be addressed in this layer are:

- Asynchronous or synchronous transmission ?
- How many start bits and stop bits ?
- Which type of modulation ? FSK (frequency shift keying) or PSK (phase shift keying)?

The interface between units and modems has to be specified in three aspects: *mechanical* (e.g. 25 pins), *electrical* (e.g. +3 to +25V represents a '0') and *functional* (describing the function of each pin).

The *data link layer* specifies the flow of blocks and error correction. By this the transmission is enhanced. For example a specification may be that a block is sent, checked and then acknowledged before a new block is sent. Hence the integrity of the data is high. One commercial example is DECnets link protocol DDCMP (Digital Data Communications Message Protocol). This can be used both for asynchronous or synchronous, serial or parallel, point-to-point or multidrop. Today special communication chips are available which support several protocols.

For the higher levels within the OSI hierarchy, standardisation is being applied, but since this deals with application, this will give more guidelines. MAP (Manufacturing Automation Protocol) which is supported by many manufacturing companies is one example of a set of protocols which covers all layers.

To illustrate the somewhat complex situation, the example of the most common type of PC connection (25-pin DCE) can be used. This is specified both by EIA(RS 232-C/D), CCITT V.24/V.28 and ISO 2110. ISO 2110 specifies the mechanical interface, V.28 the electrical interface and V.24 the functional interface. Still this is limited to layer 1.

For data networks, X.21, which defines connection and disconnection, is another example. This includes also specification of the physical layer. ISO 4903 describes the mechanical contact, X.27 and X.26 describe the electrical part and X.24 the logical functions.

3.1.6 Control system communication media and configurations

Communication media used for telephone systems include pilot wire, microwave, fibre optic links and satellites. With existing media, a communication speed of 300–2400 bps is possible and can be used for remote communication between a control centre and a substation. With the introduction of fibre optics this is rapidly increasing. Many telephone companies today offer datalinks at 19 200 bps and even in the Mbps range.

Within the substation or in a control centre the following media alternatives exist: *twisted pair*, *coaxial cable* and *optical fibre*. Twisted pair has traditionally been the most common medium for telephone communication and can also be used for datacommunication up to 10Mbps. Although comparatively cheap, it is easily exposed to disturbances and has limited capacity/speed. Coaxial cables are more immune to disturbances and can handle data rates up to 50 Mbps. For wide-band cables, which are used for cable television, up to 500 Mbps has been obtained. Optical fibre cables are the newest medium and, although they are still the most expensive, the price level is rapidly dropping. The main advantage of optical fibre is high capacity and the disturbance immunity. Data rates of 5 Gbps are possible.

The simplest type of network configuration is called *point to point*, simply a direct connection between two stations. This is however a costly and demanding way of communication when intercommunication between different units are required. The rapid development of desk-top based computers has lead to the evolution of a distributed communications system which has become known as the Local Area Network (LAN). LANs allow rapid communication between a large number of units. There are three different types of LAN configuration, or topology, possible: *star networks*, *ring networks* and *bus networks*, as shown in Figure 3.2 (overleaf).

Twisted pair and coaxial cable can be used for all three configurations. Fibre optic cable is limited to star and ring configurations since no suitable connector for bus connections has yet been developed. By using an *active starcoupler*, an optical star will have the same properties as an electrical bus. Bus and star topologies allow somewhat faster communication compared to ring. In the ring configuration, the data are transmitted from one point to the next which, in turn, resends the data to the next point. In this configuration, communication depends on all points being available, while in a bus or a star one point can fail without affecting the rest of the network. The star coupler is also a critical point which preferably should have duplicated power supplies to increase the availability.

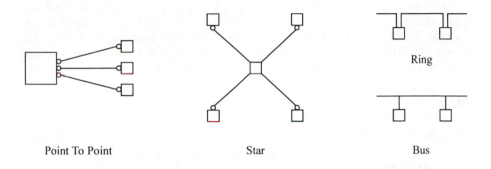

Point To Point   Star   Bus

Figure 3.2   LAN configurations

Within the environment of the substation, optical fibre cabling is always preferred due to its immunity to electromagnetic disturbances and the relative importance of the information it may be conveying. At a station level, especially in large stations with many bays, a LAN is preferable. This can either be achieved with a bus with coaxial cable or an active star with optical fibre cables. If the optical fibre star is chosen, then, in addition to the duplicated power supply mentioned above, it is also recommended that the star coupler is situated within a disturbance-protected environment.

For information exchange at object level or in smaller (distribution) stations, an optical fibre ring can be used. To increase availability it is recommended that the Object Bus Interface (OBI) has a separate power supply from the sending object which allows this object to be taken out of service without affecting the ring.

For a LAN it is important to control how the messages are sent to avoid collisions on the network. Basically two commercially available and standardised access methods exist: token ring bus with controlled token passing, which is favoured by IBM, and Ethernet CSMA/CD which is supported by DEC, Intel and Xerox. Both are included in the IEEE 802/ISO 8802 standard. IEEE 802.2 includes, for example, CRC and 802/3 CSMA/CD.

The controlled access of the token ring means that a "token" is circulated to each point on the LAN. For the CSMA/CD, a distributed random access method is used. Each point is "listening" to the network, and will not transmit data until the network is observed to be free. If a collision occurs nevertheless, the transmission is resent. Although the token ring in theory has a somewhat higher capacity compared to the Ethernet, no real preference can be given. Ethernet is, at present, supported by many manufacturers and has become an industry standard.

3.2 Fibre optic communications

3.2.1 Introduction

A comprehensive communications network is an essential requirement for the efficient operation and management of power transmission and distribution systems. Such a network is needed not only for basic telephony but also for protection signalling, control and telemetry. What suitable technology could then be employed? The choices available are as follows:

(i)   Power line carrier;
(ii) Pilot cables;
(iii) Microwave radio;
(iv) Optical fibres.

Power line carrier. This method of superimposing high frequency signals upon the phase conductors of certain high voltage transmission circuits has been successfully used for several years by the power utilities to enable the communication between associated protection schemes. Unfortunately, this communication medium suffers from restricted capacity due to the overhead line characteristics and the licensing terms because of stray radiation. Furthermore, the interfacing equipments are expensive.

Pilot cables. This technique is subject to logistic problems of traversing areas not under the control of power authorities. Additionally, for metallic pilot cables employed adjacent to high voltage power circuits, there is the possibility of interference to the limited capacity available.

Microwave radio. Usually microwave radio schemes provide a highly reliable service, but difficulties may be encountered in obtaining frequency allocations. Furthermore, they require a line-of-sight arrangement for their operation and the ideal siting of antenna towers may not be conducive to easy maintenance and/or the most direct route between terminal stations.

Fibre optics. Unlike all the previous technology, fibre optics have the following advantages:

(i)   They are not capacity-limited, i.e. they have extremely high bandwidths allowing for very large amounts of information to be carried through a very thin fibre strand.
(ii) Since the fibre is *100%* dielectric, unlike copper cables, it is not affected by ground loops, inductive pickup, cross-talk or lightning – all factors leading to noise in conventional systems. This feature is particularly important in electric utility applications where electromagnetic effects negatively impact metallic systems. Also, unlike microwave, fibre optic systems are not subject to atmospheric effects such as fading. Thus both environmental stability and low susceptibility to interference leads to a very high quality signal.
(iii) They are small and light in weight.

(iv)   There is no stray radiation from the fibres thus eliminating cross-talk and preventing unauthorised monitoring of signal traffic.

(v)   They exhibit extremely low losses which allows for long spans, thus minimising maintenance costs and the use of repeaters.

It is apparent from the foregoing that fibre optics offer significant advantages over other communication media and are therefore ideally suited to the communication requirements of the power utilities.

3.2.2 Fibre optics basics

Fibres act as waveguides in which the light is confined to the centre part, called the core, by an outer cladding, see Figure 3.3. Both core and cladding are transparent to light, but are of different refractive indices. The boundary between core and cladding appears as a partial mirror that totally reflects light rays (provided the angle of incidence is equal to or less than the angle of refraction) back into the core, thus causing it to propagate down the fibre. This is a very much simplified explanation but serves to convey the general idea.

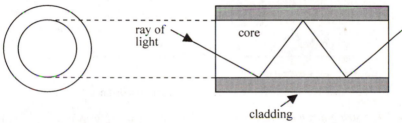

Figure 3.3   Basic fibre optic principle

3.2.2.1 Fibre types and characteristics

There are three basic fibre types commercially available: (i) step-index, (ii) graded index and (iii) single (or mono) mode.

A *step-index fibre* has only two distinct different refractive indices of the core and cladding. Because the core's index of refraction is constant, the light rays are propagated by total internal reflection at the interface between the change of refractive indices, as shown in Figure 3.4. One drawback of the step-index fibre is its limited bandwidth. This is due to two main factors, one being that light enters the fibre at a large number of angles and, as a result, some of the light rays have to travel longer distances, thus taking longer to get from one end of the fibre to the other, as shown in Figure 3.4(a). The net effect is to extend the length of a pulse and reduce its amplitude as shown in Figure 3.4(b). This is also commonly known as *dispersion*. The second limiting factor of the step-index fibre is due to the noncoherent nature of the light sources. All applicable sources available today produce light at many discrete wavelengths simultaneously. These wavelengths travel at different speeds and thus arrive at the far end displaced in time, even though they leave the light source simultaneously, thus causing signal distortion. The forementioned shortcomings of the step-index fibre limit its use to short distance transmission (usually not more than 1 km) and slow speed systems. Its primary advantage in these applications is its large acceptance angle, which allows inexpensive light sources and connectors to be used to configure a practical system.

(a)

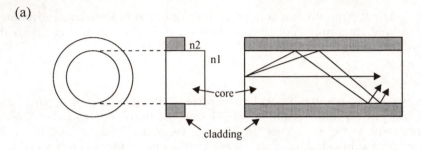

(b)

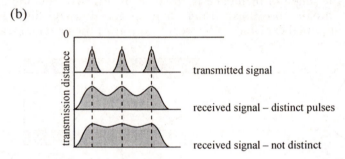

Figure 3.4   Step index multimode fibre

Graded-index fibre has been designed to overcome the foregoing problems. As shown in Figure 3.5, a graded-index fibre has its core made up of a number of layers of silica, each with a slightly different index of refraction. This causes a beam of light to change direction gradually as it travels through the different layers of the core. This tends to even out the paths of the rays entering the fibre at various angles, thus increasing the fibre's effective bandwidth. High quality graded-index fibres can have optical bandwidths approaching 1 GHz. However, they

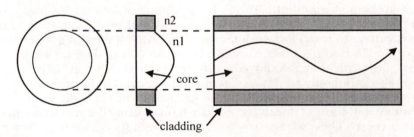

Figure 3.5   Graded-index multimode fibre

still suffer from dispersion problems to some extent. This fibre is appropriate for medium-haul medium bit rate links, typically up to 20 km at 34 Mbits/s.

As optical dispersion is caused by multiple transmission of various ray modes, *single-mode fibre* attempts to overcome this problem by only permitting one mode. This is achieved by simply reducing the cross-sectional area of the core to such a degree that only light rays with large angles of reflection will be propagated, as shown in Figure 3.6. This results in a greatly increased bandwidth and has the added advantage that because less dopant is required, the attenuation is also lower. These fibres are ideal for long-haul high bit rate use – typically 560 Mbits/s with regenerative sections of 50 km are currently possible. The drawbacks are difficulty in splicing and the small core size. The latter makes it more difficult to launch into the fibre and in fact only laser diodes (which are rather expensive) can achieve this successfully.

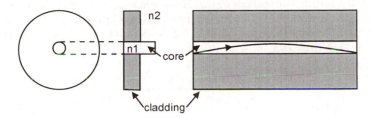

Figure 3.6   Graded-index multimode fibre

3.2.2.2 Light sources

In fibre optic communications, the source of light is invariably a semiconductor device. There are two main types, light emitting diodes *(LEDs)* and *lasers*. *LEDs* are low in cost, rugged and easy to use. *Lasers*, however, are faster, the output is more powerful and they have very small emitting areas, allowing them to launch large amounts of light into single-mode fibres.

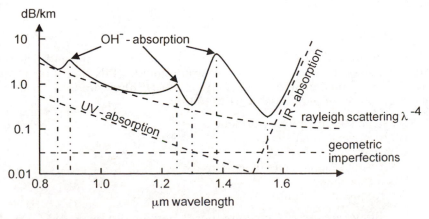

Figure 3.7   Fibre loss

Variable light is a small region in the electromagnetic spectrum covering wavelengths in the range 0.4 to 0.7 microns (μm) but the region just outside in the infra-red region can also be used up to about 1.5 μm. It has been found that the attenuation of a fibre is inversely proportional to the wavelength as shown in Figure 3.7. It is therefore advantageous to increase the wavelength of the light to reduce the attenuation. However, it is found that this general relationship does not apply throughout the frequency range and, at particular wavelengths, undesirable attenuation peaks occur (again shown in Figure 3.7). These are essentially caused by the glass going into resonance and thus significantly increasing the absorption. Attenuation minima occur at 0.8, 1.3 and 1.55 μm. Above 1.55 μm, the attenuation rises sharply due to the effect of infra-red absorption and this peak effectively places a limit on the maximum usable wavelength. Gallium arsenide-based lasers are used for producing light sources at 0.8 and 1.3 μm wavelengths. However, at around 1.5 μm wavelength, these are not very suitable and the more expensive indium phosphide has to be used for producing the laser. Table 3.1 below gives some indication of the relative costs of the devices.

Table 3.1   Relative cost of light sources

Device	Minimum cost (£)	Maximum cost (£)
0.84 μm LED	3	185
1.30 μm LED	280	500
0.84 μm laser	185	625
1.30 μm laser	1125	2185
1.30 μm laser (SM)	1375	2500

A very rough indication of device performance range is given in Figure 3.8 (opposite). This is not an accurate plot of the device performance, but an indication of the area in which each device type is expected to perform acceptably. Figure 3.9 (also opposite) shows the optical power, *P*, versus input current, *I*, of an LED and a laser diode.

3.2.2.3 Detectors

Light propagating along a fibre must be detected at the receiving end. The two main detecting devices in use are (i) *PIN* photodiode (PIN stands for p-type intrinsic n-type junction diode) and (ii) *AVALANCHE* photodiode. Some of the desirable features of photodiodes are as follows:

1   Sensitivity to the desired wavelength;
2   High conversion efficiency;
3   Low leakage;
4   Fast response to changes in light levels;
5   Small capacitance;
6   Flat frequency response.

PIN photodiodes are used for medium speed, medium distances. *AVALANCHE* photodiodes are used for very high speed digital systems.

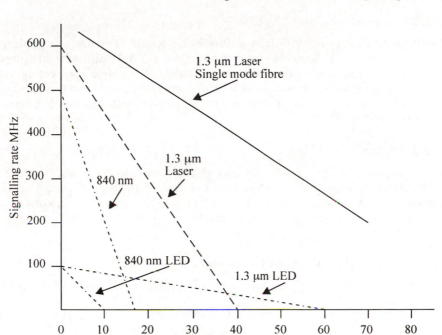

Figure 3.8   Device performance range

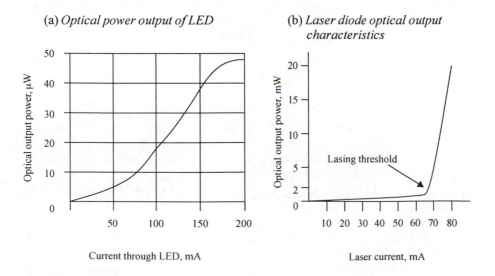

Figure 3.9   Optical power output characteristics of LEDs and lasers

3.2.2.4 Connectors

Connectors are the weak link in a fibre optic system. For low speed, short distance applications, low cost plastic connectors, e.g. AMP optimate, are quite adequate. When better performance is needed, *SMA* style connectors are employed. In applications where the maximum performance must be squeezed out of the system, very high performance connectors with losses in the range 0.5–1 dB are used.

3.2.2.5 Splices

Different pieces of fibres have to be joined. The four basic methods are fusion, mechanical, epoxy and elastomeric. Of these, fusion splicing is the most widely used as it keeps the attenuation to a minimum. In this method, a high intensity electric arc is generated which melts the fibres together forming a relatively strong interface.

3.2.2.6 Lightwave modulation techniques

There are two basic techniques for transmitting information over a fibre optical link: analogue and digital.

Analogue modulation. The simplest analogue technique directly intensity modulates the optical source with the message signal, see Figure 3.10. The advantage of this approach is its simplicity and low cost. The disadvantage is nonlinearities in the optical sources which severely limit the system performance, restricting both transmission distance and signal-to-noise performance. Presently, most practical analogue fibre-optic systems employ modulation techniques that reduce the effects that the device nonlinearities have on transmission performance. The commonly employed techniques are frequency modulation and pulse frequency modulation and these can improve signal-to-noise ratios over direct intensity method by 10–15 dB. The disadvantage is the complexity and high cost.

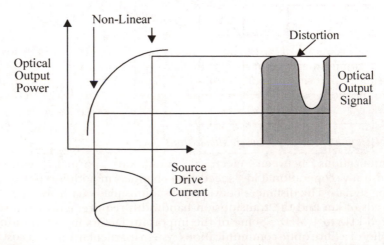

Figure 3.10 Distortion in analog modulation of LED

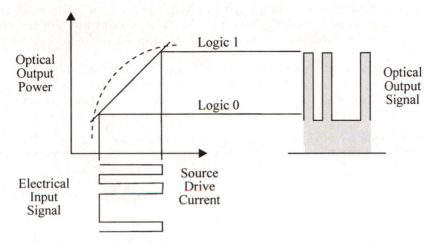

Figure 3.11 Digital modulation of optical source

Digital modulation techniques are the simplest to implement and result in a system that can transmit over great distances without degrading the message signal's performance. Pulse code modulation (PCM) signals are used to intensity modulate the optical source directly as shown in Figure 3.11. Because the only information transmitted is the presence or absence of pulses, nonlinearities in the system have no effect on the transmission quality. Providing the receiver can distinguish between a pulse and a no-pulse condition (i.e. there are no bit errors), no noise or distortion will be added to the transmitted information.

3.2.3 Communications applications in power systems

So far we have discussed briefly some of the physics associated with optical fibres. The following section describes how the various optical factors can be taken into account to provide a suitable communication system, for example to the requirements of the a power utility. There are three main areas of interest: communication between substations usually over long distances, communication inside the substation for local networking over short distances and connection of station communications equipment with an external network.

3.2.3.1 Communication between substations

Communications between electrical substations normally consist of administrative/operational low speed data channels in addition to protective relaying signals. The distances between substations may vary from a few km to hundreds of km and the transmission bandwidth requirements from several hundred kHz to 1 MHz . Some of the important factors in determining the feasibility of a fibre optic communications system, particularly from a cost point of view, are transmission distances, capacity and the use of repeaters in the line.

To ensure that there are sufficient operational optical margins, it is essential that a certain amount of budgeting is undertaken to cater not only for the intrinsic cable losses but also for connector losses, degradation of performance, etc. This allocation of losses is called an *optical budget.*

Knowing the attenuation coefficient of the optical fibre (usually 0.5 dB/km for a 1.3 µm system) and using single mode propagation to reduce signal dispersion allows regenerator section lengths to be estimated and maximised. Table 3.2 shows the budget for a single mode 140 Mbits/s system.

Table 3.2   30 km 140 Mbits/s single mode optical fibre system

1 Optical path loss		
Total fibre length	30.5 km	
Fibre loss at 1.3 µm	0.5 dB/km	15.2 dB total
Average splice loss	0.2 dB	
No of splices	14	2.8 dB total
Total connector loss		3.0 dB
Total path loss	*21 dB*	
2 Power budget		
Mean launched power	-10 dBm	
Receiver sensitivity	-42 dB	
Available power ratio	*32 dB*	
Path loss	21 dB	
Operating margin	*11 dB*	

Another example of a budget for maximising span between repeaters is shown in Table 3.3.

Table 3.3   565 Mbits/s single mode fibre

	1.3 µm	1.5 µm
Emitted power	0.0	0.0 dBm
Connector losses	2.0	2.0 dB
Equipment margin	3.0	3.0 dB
Received sensitivity	-32.0	-30.0 dB
Allowed loss in waveguide	27.0	25.0 dB
Fibre attenuation	0.5	0.3 dB/km
Splicing losses	0.2	0.15 dB
Link margin	0.2	0.15 dB
Unitary length of waveguide		
Maximum system span	27/0.9 = 30.0 km	25/0.6 = 42.0 km
Distance limit due to dispersion of pulses	300.0 km	50.0 km

Such budgets must be drawn up for any system. The main variables between systems are the fibre loss, the mean power launched and the receiver sensitivity. The light sources are either high radiance LEDs or laser diodes and the detectors either AVALANCHE photodiodes or PIN detectors. As a general rule, LEDs are used with graded index fibre and lasers with single mode.

3.2.3.2 Types of optical cables suitable for use on power lines

Externally attached optical cable. Several systems have been developed for attaching optical cables outside overhead line conductors. Perhaps the most successful is the UK scheme where the optical cable is wrapped around a conductor. This avoids the aerodynamic lift associated with parallel cables and removes the need for any form of clip, which may cause electrochemical corrosion problems. The advantage of wrapped cable over alternative systems is that it offers a potentially inexpensive cable with very low installation cost. Ease of installation is particularly beneficial in difficult terrain. Protective scaffolding is not needed to cross obstructions such as low voltage power lines, motorways and railway, so disruption is minimised. Cable can be wrapped onto an earthwire without the need to isolate adjacent circuits. In addition, wrapped cable is suitable for both low voltage distribution lines and high voltage transmission lines. Distribution lines penetrate much further into the centres of population between which wide band communication is required. The main disadvantage of small externally attached cables is that they can be vulnerable to mechanical damage.

Optical composite conductors for overhead head line systems. Optical composite conductor has been under development in the UK since about 1976. The first UK trial length was installed in mid-1979 and a full scale system trial on a 400 kV line has been in operation at 34 Mbits/s since 1982. Extensive testing, covering all aspects of mechanical, electrical, handling, fault and lightning performance has shown the concept to be sound. Optical composite conductor has been in routine engineering use in the UK since about 1984. An optical cable of the loose-buffered type, usually surrounded by a protective metal tube, forms the centre of stranded overhead line conductor which may be used as a phase or earth conductor. Around the aluminium alloy tube are stranded wires of the required diameter, material and number; Figure 3.12 is a typical example.

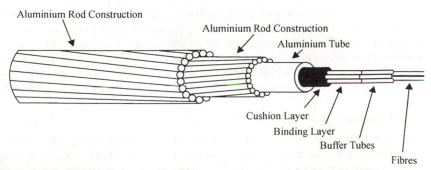

Figure 3.12 Composite overhead cable

Aluminium alloy, galvanised steel of various strengths and aluminium-clad steel wires may be stranded either with one type or a mixture of several types of wire to match the characteristics of the line conductor. Protection against corrosion and water ingress is provided by filling the interstices between the wires and any space within the closed aluminium alloy tube with a high melting point grease. Composite conductors must match their conventional counterparts in terms of strength, weight per unit length, sag, diameter, conductance and in their behaviour to temperature variations, ice and wind loading. This can be achieved quite readily for conductors larger than about 14 mm in diameter although smaller conductors can present special problems.

Composite conductor is supplied in lengths of up to 4 km and may be handled much as conventional conductor. Composite conductor must remain uncut at intermediate tension towers with a loop between dead-ends forming a jumper. Conductor fittings must be chosen with care to prevent radial distortion of the conductor.

Self-supporting, all dielectric, aerial optical cable. This type of cable is intended to be strung on existing overhead line supports. It must therefore meet the mechanical requirements of an overhead line conductor. The same ice and wind loading, ground clearance, fittings type testing etc will apply.

An additional constraint for metal-free cables is that of clearance from phase conductors during high winds transverse to the line. Experience has shown that metal-free cables blow out a great deal further, and their response time to changing wind conditions is a lot shorter than standard conductors.

At the higher transmission voltages, a design problem to which no solution has yet been proved satisfactory is that of sheath degradation caused by exposure to electric fields. Experience in Europe has shown that installations on lines of up to 110 kV have been satisfactory but problems have been experienced at 220 kV. Considerable research effort is being expended to find a sheath material which is resistant to this type of attack.

3.2.3.3 Communication inside the substation

Fibre optic applications are associated with local networks inside the substation for SCADA and protection purposes. In these applications, transmission distances are no longer than 1 km and information transmitted is in the low frequency range (less than 1 MHz). Here the fibres are used mainly for their dielectric properties.

Data transmission inside electrical stations represents an ideal application for short distance fibre optic links. Due to the low signal levels normally found in conventional metallic data transmission systems, they have not always been suitable for operation under high electromagnetic interference very often found in the substation environment. Fibre data transmission links are used for connecting data acquisition equipment with the processing unit in SCADA systems.

3.2.3.4 Communication equipment with external network

These can be considered as short distance telecommunication links used to interconnect the station communication room with an external communication network that may be private or public. These links are normally between 1 and 5 km long and the main characteristic of the fibres used is their immunity and isolation against high voltage and current surges in the station. In any case, the use of fibres eliminates the problem of potential earth differences between the station and the network terminal.

3.3 Further reading

1   SLOANE, A.: 'Computer communications: principles and business applications' (McGraw-Hill International, 1994)
2   DUUREN, J. van, KASTELEIN, P. and SCHOUTE, F .: 'Telecommunications networks and services' (Addison-Wesley, 1992)
3   HALSALL, F.: 'Data communications, computer networks and open systems' (Addison-Wesley, 1992)
4   LILLY, C. J. and WALKER, S. D.: 'The design and performance of digital optical fibre systems', *Proceedings of the Radio and Electronic Engineer*, **54**, (4), **1984, pp. 179–191**
5   ELLEN, W. N. and PLAATS, J.: 'Fundamentals of optical fibre communications' (Prentice Hall International series on optoelectronics, 1991)
6   HOSS, J. R.: 'Fibre optic communication design handbook' (Prentice Hall International, 1990)
7   DUNLOP, J. and SMITH, D. G.: 'Telecommunications engineering, 2nd edition' (Chapman and Hall, 1989)

Chapter 4
Numeric protection
Dr P. J. Moore

Introduction

This chapter describes, in some detail, the hardware, operating principles, software design procedures and testing of contemporary numeric relays and fault recorders.

Following on from Chapter 1, which introduced the technology of digital techniques, Section 4.1 draws on this knowledge to give a general overview of the internal hardware structure of a numeric relay. Section 4.2 introduces the operating principles of the main types of numeric protection currently available. In general the algorithms consist of a stage of signal formation, e.g. impedance measurement in the case of a distance relay, followed by a stage of fault evaluation. It is assumed that the sampled input signals to the distance and directional comparison algorithms have already been digitally filtered. Section 4.3 reviews the current trends in numeric fault locating techniques. Both Sections 4.2 and 4.3 are concerned with *algorithms* – a description of the processing required to solve a given problem. Since algorithms are ultimately converted into computer software, an overview of this conversion procedure is described in Section 4.4. Microprocessor technology has also improved the methods by which relays may be tested and this is described in Section 4.5. Finally, some examples of commercially available numeric relays are given in Section 4.6.

4.1 Numeric relay hardware

4.1.1 Typical relay hardware structure

Figure 4.1 (overleaf) shows the general hardware outline of a numeric protection relay. Relaying voltages at 110 V or 50 V and currents, at 5 A or 1 A, are first passed through isolation transformers. Since analogue to digital conversion is usually performed on voltages, the current signals are converted to representative voltage signals by, for example, passing the current through a known resistance value. All the signals are then filtered using very simple analogue anti-aliasing filters (see Section 2.2). Since ADCs are expensive it is common to find only one used in a digital relay, thus an analogue multiplexer, under microprocessor control, is used sequentially to select the required signal into the ADC. Because the ADC takes a finite conversion time, typically 25 µs, it is necessary to hold the incoming signal for the duration of the conversion; this is achieved with the sample and hold amplifier. Having been converted by the ADC, the signals can now be manipulated by the microprocessor. It is common to find more than one microprocessor used, e.g. a TMS320 for executing the relay algorithm and a 80186 for the scheme logic. The relaying program will be located in the read only memory (ROM), and the random access memory (RAM) will be used for storing sampled quantities and intermediate products in the relaying algorithm. Relay settings are stored in the electrically erasable programmable read only memory (E^{2}PROM).

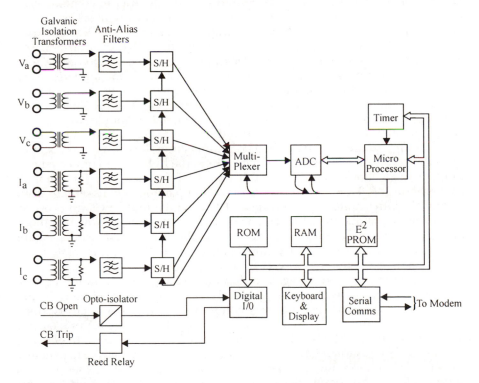

Figure 4.1   Typical numeric relay hardware

Relays are powered from the station batteries which are typically 50 V. Since the battery voltage is prone to variation depending on its state of charge, a power supply is incorporated in the relay to provide regulated, constant power rails for the relay electronics. These are typically ±5V and ±12V. Switched mode power supplies are normally used in relays since they are more efficient, dissipating less power, and can work with a wider variation in supply voltage than more conventional series regulator types. In addition, switched mode power supplies also allow isolation between the station batteries and the relay electronics.

4.1.2 Relay interfaces

Communication with a relay is necessary for three reasons. First, a facility must exist for programming settings into the relay, secondly, unit type relays must communicate with their counterparts and, thirdly, the relay must issue trip and alarm signals under onerous conditions.

Unlike electromechanical or static relays, numeric relays have few or no case-mounted controls for adjusting settings. Settings are normally associated with the relaying program and have consequently to be entered into the software. Hence, some form of interface is required to allow the user to communicate with the relay. This form of communication can usually occur on two levels. Firstly, it is common for contemporary numeric relays to have liquid crystal displays (LCD) and keypads incorporated on the relay front panel, see Figure 4.1. To enter settings, the user manipulates the keypad to display and alter numbers appearing on the LCD display. Note that some numeric relays do not use K factors to specify the settings, instead actual values, e.g. sequence impedance values, are used. However, manufacturers that do use K factors are now supplying users with programs, to run on personal computers, which allow easy computation of the K factors. Secondly, a visual display unit (VDU) may be connected to the relay via a serial communications link. This may occur in the substation or it may occur remotely if the serial link can be connected to a modem which allows transmission of the serial information over, for example, a telephone line. Thus, it is possible to dial up a relay and alter its settings from a control centre. Communication via a VDU is similar to the LCD and keypad approach except that the VDU screen and keyboard are used. Note that there are many different specifications, or protocols, for serial data transmission; the most likely type for remote relay interrogation is the RS232 protocol. Further details on digital communications can be found in Chapter 3.

Unit-type relays, such as digital differential relays, communicate digitally with their counterparts using a form of serial communication. Although some types are designed to use existing communication media such as voice frequency communication links, others use a completely digital approach and are designed for 64 kbit/s pulse code modulated (PCM) systems. In general, digital communication is superior to analogue communication due to its ability to check for, and to some extent correct, errors in transmission. The circuitry required for digital communication is integrated into the relay hardware and is under control of the relay microprocessor, see Figure 4.1.

Digital relays need some method of issuing trip signals and alarm signals. Since these signals are essentially binary, it is relatively simple to decode some part of the microprocessor address space for this use. This occurs in the block marked *digital output* in Figure 4.1. Ironically, despite the advanced technology found within a digital relay, the trip and alarm signals are commonly connected to the outside world via electromechanical reed relays.

4.1.3 Relay operating environment

The substation is an electromagnetically hostile environment for a numeric protection relay due to its close proximity to transmission lines, isolators and circuit breakers. Under fault and switching conditions it is essential that external noise does not enter the relay and prevent normal relay operation. Disturbances in this category are classed as electromagnetic interference (EMI).

There are essentially two causes of EMI in substations: first, as a result of switching operations and line surges, which may also couple onto low voltage relay input signals and, secondly, due to external causes such as lightning strikes to power system equipment and spurious radio interference. It should be stressed that, because of the high-speed operation of the microprocessor, numeric relays are especially prone to EMI effects. Thus, it is imperative to ensure that a numeric relay has electromagnetic compatibility (EMC) with its environment.

To ensure digital relays meet EMC requirements, a variety of approaches are taken. These range from encasing all relay electronics within a steel faraday cage to isolating all connections to the relay galvanically. Increasingly, relays are tested to rigorous IEC specifications to ensure greater quality control for the user. It is worth noting that electromechanical relays were largely unaffected by EMI and so introducing digital technology into the substation environment has created some problems as well as benefits. A typical utility specification for a numeric relay in respect of electrical environment, insulation and EMC is given in the Appendix, Section 4.9, which mentions some of the relevant IEC specifications.

4.2 Numeric relay algorithms

4.2.1 Overcurrent relays

4.2.1.1 Introduction

Overcurrent relays are the least complex type of relay to be implemented by numeric means. Due to the relative slow operation of an overcurrent relay compared with, say, a distance relay, there is little performance benefit to be gained from a numeric implementation.

The main benefit of numeric overcurrent relays is lower cost and the ability to provide a full range of characteristics in one product, the required characteristic being selected by switches on the relay front panel.

With respect to the hardware of a numeric overcurrent relay, the outline is essentially as given in Figure 4.1 excepting the following points:

(a) Only one current input is provided per relay;
(b) The processing requirements are far less demanding for an overcurrent relay than the other types described in this chapter;
(c) By virtue of (a) and (b), it is possible to integrate the data acquisition and microcomputer components within one chip, thus reducing printed circuit board space within the relay;
(d) Some relays use a bridge rectifier on the current input to avoid using a bipolar analogue to digital converter.

4.2.1.2   Operation

The operation of a typical numeric overcurrent relay will be described. The signals within the relay are shown in Figure 4.2. The current into the relay is firstly rectified and then passed through a resistor network, selected by switches on the front panel, to provide a voltage proportional to the incoming current. The switches are the equivalent of the *plug setting multiplier* found in electromechanical overcurrent relays and serve to scale the input current.

The scaling is such that, irrespective of the current setting, input current at the setting level will produce the same internal voltage in the relay. This voltage is then digitised by an analogue to digital converter. Sequential samples are then compared to find the *peak values* of the rectified sine wave. These peak values are stored in *peak registers* within the microprocessor; four peak registers are used to store the preceding four peak values. Every time a new peak value is added, all the peak registers are compared to find the *highest peak value* over the last four peaks. The highest peak value to then referred to a *look-up table* (a table of coefficients stored in memory) which produces an *increment number*. The increment number depends on the highest peak value and is seen to change in Figure 4.2 from Y to Z as the input current to the relay increases in magnitude.

In addition, a timer and counter are used to count the number of successive periods of 3.2 ms between peaks. When a peak is detected, the counter is reset and begins to count from zero. The number of 3.2 ms counts is then multiplied by the increment number to form a value which is added to the *trip time register*, this also occurs every time a peak is detected. If the trip time register exceeds a value K, then the relay will trip. The value of K is effectively the *time setting multiplier* of the relay. The characteristic of the relay, i.e. whether it is inverse, very inverse, instantaneous etc., is determined by the look-up table which sets the increment number. Using this method, all types of overcurrent characteristic can be obtained.

The overcurrent technique described above is suitable for implementation by a standard 8 bit microprocessor. With more powerful microprocessors available, it is likely that future relays will incorporate Fourier techniques, i.e. the Discrete Fourier Transform, thus allowing digital filtering of the input current to ensure that only power system frequency sinusoids are processed.

4.2.2   Distance relays

4.2.2.1   Introduction

Distance relays were one of the first protective devices to be considered for numeric implementation yet, ironically, they are currently the least mature of all numeric relay types in terms of commercial development, with products only recently emerging into the market place. The most probable reason for this is the relative complexity of a numeric distance relay when compared with other types described in this book. Thus the recent arrival of powerful microprocessors to enable practical commercial designs was instrumental to this progress.

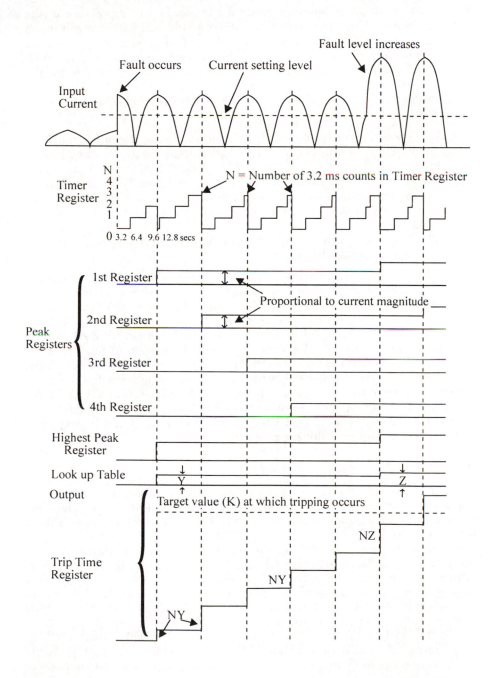

Figure 4.2   Signals showing the operation of a numeric overcurrent relay

Numeric distance relays differ from more conventional static types in that they calculate an actual numeric value for the apparent impedance at the relaying point. This impedance is subsequently compared against an impedance plane-based characteristic in order to make a relaying decision. In static distance protection, e.g. by using a block-average comparator, the relay functions by directly combining the voltage and current inputs in the comparator to form the relaying decision.

Although the end result is the same in either case, the following advantages apply to the numeric relay:

(a) Since both the phase and amplitude information of the input signals are used, the security of the relay is higher than if only, say, the phase information is used;
(b) Any shape of characteristic can be easily programmed into the relay;
(c) Zones of protection are easily incorporated since, once the impedance has been calculated, extra zones may be added with little processing penalty;
(d) The characteristic may be set with ohmic values, not K values, thus simplifying commissioning.

Many different approaches to impedance measurement have been attempted. One approach is an algorithm based upon Fourier techniques. For example, if the discrete Fourier transform (DFT) is applied to the voltage and current samples, the resulting phasors can be combined by complex division to form resistance and reactance quantities. The large majority of algorithms, however, are based upon solving the first order differential equation of the line; such algorithms assume a series resistance and inductance model of the line – shunt capacitance is neglected. Since algorithms of this nature assume that the input signals are sinusoidal, it is necessary to prefilter the signals with a digital filter to ensure that only the power system frequency component is processed. By comparison, algorithms using DFT techniques do not need prefiltering since the DFT has an inherent filtering property.

4.2.2.2 *Impedance-measuring algorithm*

The current trend in numeric distance relays is to use an impedance measuring algorithm based on the first order differential line equation. Whilst there are a plethora of techniques to solve this equation, only one technique will be described here.

Assuming the transmission line to be modelled by a series resistance, R, and inductance, L, as shown in Figure 4.3, the first order differential line equation is given as:

$$v_r = Ri_r + L\frac{di_r}{dt} \tag{4.1}$$

where the relaying voltage and current are v_r and i_r respectively. It is more convenient to abbreviate the current derivative of Equation 4.1 and so we will write:

$$v_r = Ri_r + Li'_r \quad \text{where} \quad i'_r = \frac{di_r}{dt} \tag{4.2}$$

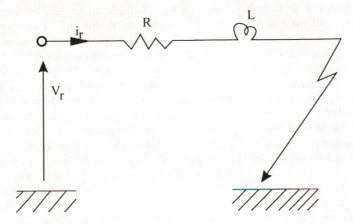

Figure 4.3   Distance relay model of transmission line

To calculate the line impedance, the relay must solve Equation 4.2 for R and L. However, to solve for two unknowns, two equations are required and thus two separate solutions to Equation 4.2 must be found. We will later describe how this is achieved, but, for now, it will be assumed that two sets of relay voltages, currents and current derivatives are available. These will be written as:

$$v_{r1} = Ri_{r1} + Li'_{r1} \tag{4.3}$$

and

$$v_{r2} = Ri_{r2} + Li'_{r2} \tag{4.4}$$

where the subscripts 1 and 2 distinguish between the two solutions. Equations 4.3 and 4.4 may be combined to form two new equations:

$$R = \frac{v_{r1} i'_{r2} - v_{r2} i'_{r1}}{i_{r1} i'_{r2} - i_{r2} i'_{r1}} \tag{4.5}$$

and

$$L = \frac{v_{r2} i_{r1} - v_{r1} i_{r2}}{i_{r1} i'_{r2} - 1_{r2} i'_{r1}} \tag{4.6}$$

For convenience, the denominator of Equations 4.5 and 4.6 will be replaced with:

$$D = i_{r1} i'_{r2} - i_{r2} i'_{r1} \tag{4.7}$$

Thus, the apparent resistance and reactance presented to the relay may be calculated as follows:

$$R \times D = v_{r1} i'_{r2} - v_{r2} i'_{r1} \tag{4.8}$$

and

$$X \times D = \omega LD = \omega(v_{r2} i_r - v_{r1} i_{r2}) \tag{4.9}$$

where ω is the power system frequency and X is reactance.

Note that the D term has been cross-multiplied in Equations 4.8 and 4.9. This is to avoid the need for digital division which will put a large burden on the relay processor. In the subsequent fault evaluation process where the measured impedance is compared with the reach point values, these values are first premultiplied by D before the comparison is made. It will be appreciated that since the power system frequency value, ω, is arbitrarily specified, the algorithm is not sensitive to changes in actual power system frequency.

It was earlier assumed that two solutions to Equation 4.2 may be found. The simplest method of achieving this is to space the solutions in time. Figure 4.4 shows this approach where the derivations of v_{r1}, v_{r2}, i_{r1} and i_{r2} are made from different points on the sampled waveforms; a time difference, t_{diff}, is seen to exist between solutions 1 and 2. The choice of the value of t_{diff} is a balance between ADC accuracy and relay operating time. Making t_{diff} too small leaves the ADC unable to resolve the differences between the two solutions, whilst making t_{diff} too large will lead to an unnecessarily long operating time since the post-fault impedance values will only be correct when both solutions are derived from post-fault data. In practice a value of t_{diff} of 1.5 ms is ideal when a 14 bit ADC and 4 kHz sampling frequency is used.

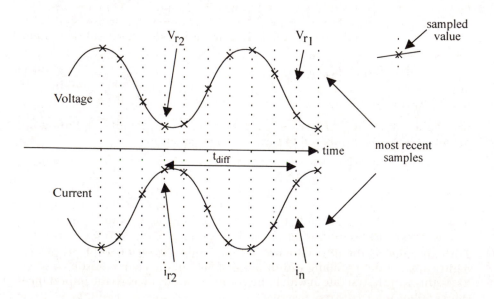

Figure 4.4   Time spaced solutions

The current derivatives are calculated by assuming piecewise linearity as shown in Figure 4.5. Note that this involves using a sample in advance of the point in interest, thus the values of v_{r1} and i_{r1} in Figure 4.4 are the sampled values taken before the most recent sample points.

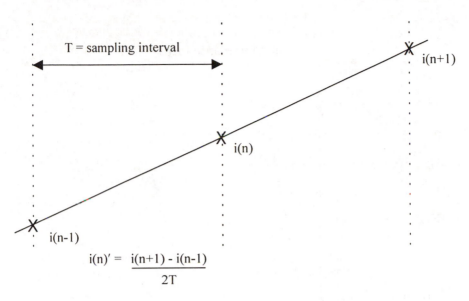

$$i(n)' = \frac{i(n+1) - i(n-1)}{2T}$$

Figure 4.5   Calculation of current derivative

4.2.2.3 Fault evaluation

In order to ascertain whether or not a fault exists on the line, the calculated impedance needs to be compared with an impedance characteristic. Figure 4.6 shows a quadrilateral characteristic used for the fault evaluation. In general, any shape of characteristic could be realised with a digital distance relay since the shape is defined in the software of the relay. The occurrence of a single calculated impedance value falling within the characteristic is not a reliable indication of a fault since the relay filtering process is not ideal. To ensure high relay integrity in evaluating a fault, a counting strategy is employed which weights the impedance samples depending upon their proximity to the reach point. The characteristic has associated with it a counter which, prior to the fault is set at 0. When the first calculated impedance value enters the characteristic, the counter is advanced by a value of δ, which is set according to the position of the impedance value within the characteristic as shown in Figure 4.6. For values falling within the first 80% of the characteristic, δ is set to +9; this is the fastest rate at which the counter can advance. As the impedance values fall nearer to the reactive reach point, δ decreases to +4 and then +1 for faults within 10% of the reach point. The positive δ values in zone of the characteristic are reflected negatively on the out-of-zone side beyond the reactive reach point. This helps to promote correct reach point operation. The resistive reach of the characteristic is not graded in a similar fashion since the presence of fault resistance can greatly influence the measured value. If the counter value reaches 45 then the relay trips. If the impedance falls anywhere outside the characteristic other than beyond the reactive reach point, δ is set to -9. The counter is never allowed to go negative.

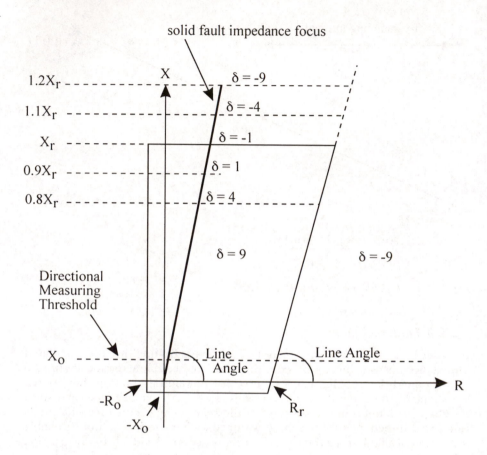

Figure 4.6   Quadrilateral characteristic and counting strategy

4.2.2.4 Directionality

Faults close to the relay location may result in the measured voltage falling to zero thus leaving the relay unable to resolve whether the fault is in the forward or reverse direction. To give the relay directionality, a reactance is calculated using pre-fault voltage samples and post-fault current samples. This is achieved by delaying the sampled voltage values in memory for a whole number of power system cycles; the number chosen will reflect the number of cycles for which the relay will remain directionally stable. The resulting reactance, referred to as the *directional reactance*, will be positive for faults in the forward direction and negative for faults in the reverse direction. Figure 4.7 shows the arrangement for calculating the directional reactance using a one cycle memory for the voltage.

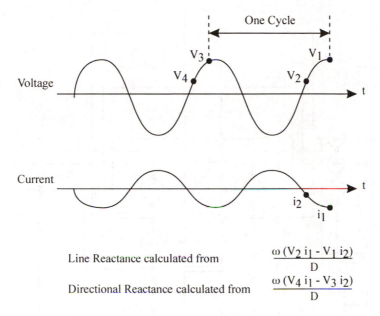

Line Reactance calculated from $\dfrac{\omega (V_2 i_1 - V_1 i_2)}{D}$

Directional Reactance calculated from $\dfrac{\omega (V_4 i_1 - V_3 i_2)}{D}$

Figure 4.7   Device performance range

Thus, an additional constraint for the counter described in Section 4.2.2.3 to be advanced is that the directional reactance is greater than the directional measuring threshold of Figure 4.6. If the directional reactance is less than this threshold, no relay operation can occur.

4.2.2.5 Relay elements

In common with conventional distance relays, a numeric distance relay uses phase and earth fault elements to detect faults. In a single zone relay, there will be 3 elements to detect faults involving earth paths, and 3 elements to detect faults between phases. Each of these elements will calculate a separate impedance and perform a fault evaluation. The difference between the elements is the voltage and current samples that are provided for the impedance calculation. Figure 4.8 shows the signal flow of the sampled phase voltages and line currents. The earth elements are residually compensated by adding a fraction of the residual current to the line currents. For overhead lines the residual compensation factor is usually approximated by a scalar value (i.e. the phase angle is zero). However, it is relatively simple to use complex residual compensation factors [1].

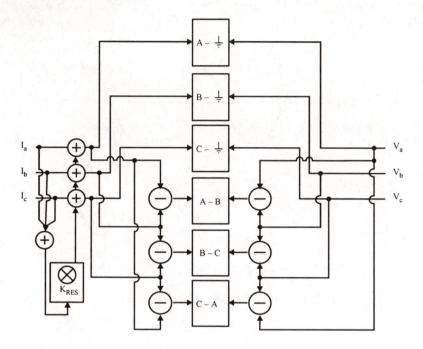

Figure 4.8   Signal flow path for distance relay elements

4.2.3 Directional comparison relays

4.2.3.1 Introduction

The use of directional relays, positioned at either end of a transmission line, to compare whether a fault is internal or external to the protected line, is widespread. Although this type of protection has been superseded by phase comparison relays, the advent of numeric protection relays has seen the development of a new form of directional comparison relay based upon *superimposed components.* The advantage of this approach is that correct directional decisions can be made even if the minimum fault current is less than the maximum load current.

Similar to the numeric relay algorithm, the superimposed component directional comparison relay algorithm assumes that the input signals are sinusoidal and so it is necessary to filter the signals digitally prior to relay processing. A typical sampling frequency for a numeric directional comparison relay is 3 kHz.

4.2.3.2 Superimposed components

A faulted power system may be analysed as though it is composed of two networks added together. These networks are the original pre-fault state of the power system and, secondly, a superimposed system where the sources are shorted out and a voltage source at the fault point is inserted. Voltages or currents existing in the superimposed system are referred to as 'superimposed components'. This is shown in Figure 4.9 in which a faulted two-ended transmission line model is decomposed into the two described networks. In this example, a solidly earthed single phase to earth fault occurs at the mid point of the line. Clearly there is zero voltage at the fault point in the faulted network, and hence the fault point voltage source in the superimposed network is set to a value equal and opposite to the voltage existing at the mid point of the line prior to the fault. This ensures that, when the pre-fault and superimposed networks are added, zero voltage at the fault point will occur in the faulted network.

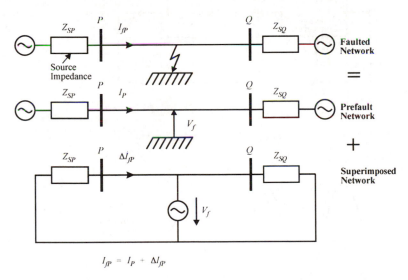

$$I_{fP} = I_P + \Delta I_{fP}$$

Figure 4.9   Superimposition of networks to form a faulted system network

Since the superimposed network is immune from the effects of pre-fault loading, it is advantageous to base the operating principle of a protection relay upon superimposed components since the relay operation will also be immune to pre-fault loading. In order to form superimposed components, for example voltage, it is necessary to subtract the voltage existing prior to the fault from the voltage measured during the fault, and similarly for current. Although this sounds simple in principle, small variations in power system frequency make it a difficult task.

The superimposed components may be formed as shown in Figure 4.10 where sampled voltage values are delayed in memory for one cycle; the superimposed voltage is then formed by subtracting the delayed sample from the most recent voltage sample. This arrangement assumes that the sampling frequency is an integer multiple of the power system frequency and so there will be a whole number of samples taken per power system cycle. If this is so, then the one cycle delay can be implemented by delaying the samples in a fixed number of memory locations. If, say, the system frequency were to drop slightly, then a power system cycle would be slightly longer in time and there will now be more samples needed to represent one cycle. Furthermore, it is unlikely that a whole number of samples would be needed per cycle. This will lead to an error in the calculation of the superimposed components due to the fixed number of memory locations used for the one cycle delay. To overcome this, a *frequency tracking* circuit, usually a phase-locked loop, constantly adjusts the sampling frequency to ensure that one cycle of measured voltage samples

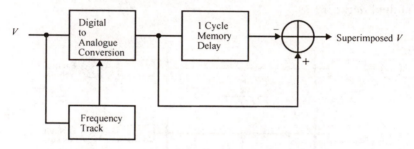

Figure 4.10 Digital extraction of superimposed voltage

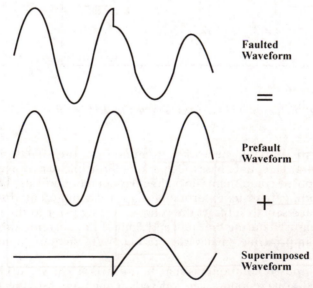

Figure 4.11 Superimpostion of waveforms

always occupies a fixed number of samples. An example of a faulted waveform and its superimposed component is shown in Figure 4.11.

Using this technique the superimposed component will only be present for exactly one cycle. Since directional comparison relays can give fault direction decisions in only a few milliseconds, this limitation on the superimposed component signal lifetime is not important.

4.2.3.3 Directional measurement

Consider the superimposed network of Figure 4.11 which also shows two directional relays located at busbars P and Q respectively, which are both 'looking' in the same direction towards busbar R. The superimposed current and voltage signals measured by the relay at P, for which fault will appear in the forward direction, are related by the source impedance behind the relay location:

$$\Delta V_p = -\Delta I_p Z_s \qquad (4.10)$$

However, for the relay at Q, for which the fault will appear in the reverse direction, the relationship is:

$$\Delta V_q = +\Delta I_p Z_{st} \qquad (4.11)$$

where Z_{st} is the impedance of the line between Q and R plus the impedance of the source beyond R.

This basic change in sign is the basis for the directional decision. In practice the magnitudes of Z_s and Z_{st} are relatively unimportant since it is only the polarity of the signals which is used in the directional determination. However the phase angle is important since, in order to be able to compare the ΔV and ΔI signals conveniently, it is necessary to adjust these quantities so that the signals appear in phase during fault conditions. In practice an assumed phase angle of 78° adequately represents the likely variation in source and line impedances.

Forming cophasor signals for the directional decision process can be achieved in two ways. Firstly, ΔI can be multiplied by a replica impedance representative of the likely values of Z_s and Z_{st}, thus making ΔI in phase with ΔV. Secondly, an all digital approach, the superimposed voltage is delayed by an angle ϕ, usually 78°, giving $\Delta V \angle \phi$; this can be conveniently performed by a

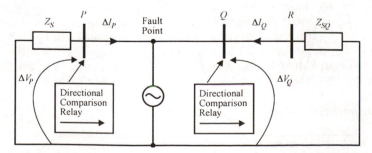

Figure 4.12 Superimposed network with only forward looking directional comparison relays

digital filter having the required phase shift property. In addition, some scaling of the magnitudes of ΔV and ΔI is necessary.

Hence, with reference to Figure 4.12, in general for a forward fault:

$$\Delta I = \frac{-\Delta V \angle \phi}{|Z_s|} \tag{4.12}$$

where ΔI and $\Delta V \angle \phi$ are of opposite polarity, and for a reverse fault:

$$\Delta I = \frac{+\Delta V \angle \phi}{|Z_{st}|} \tag{4.13}$$

where ΔI and $\Delta V \angle \phi$ are the same polarity.

The comparison is then performed by using the following mixed signals:

$$S_1 = |\Delta V \angle \phi + \Delta I| \tag{4.14}$$

$$S_2 = |\Delta V \angle \phi - \Delta I| \tag{4.15}$$

where $(S_2 - S_1)$ is positive for forward faults and negative for reverse faults.

In order to make a directional decision, the value of $(S_2 - S_1)$ is compared with two separate threshold levels, one positive for forward faults and one negative for reverse faults. Similar to the case of the numeric distance relay, the single occurrence of $(S_2 - S_1)$ exceeding the threshold level is not a good indication of directionality and a counting strategy is employed to smooth out transient data. The counting strategy essentially ensures that $(S_2 - S_1)$ exceeds the relevant threshold three consecutive times before a directional decision is made.

4.2.3.4 Directional elements

Three separate elements are used to detects faults on all phases. Each element is provided with either a–b, b–c or c–a superimposed voltage and current signals. Thus, each element can detect a specific phase-to-phase fault. Phase-to-earth faults are detected by two of the directional elements picking up.

4.2.3.5 Application

In use, directional comparison relays are situated at each end of a transmission line and arranged so that their forward directions are pointing at each other. A communication channel between the relays is necessary to convey the directional information to and from each end. This channel may operate in permissive or blocking intertripping modes as described elsewhere. In addition, it is common for numeric directional comparison relays also to have an underreaching instantaneous independent mode of operation which gives protection of approximately 30% of the line in the forward direction. This independent zone can be a simplified form of a distance relay, or a directional overcurrent relay, enabled when a forward decision is reached.

4.2.4 Differential relays

4.2.4.1 Current measurement

A current differential relay performs a Kirchoff Law current summation of the currents entering a multi-ended circuit. By comparison of the residual components of the summation it is possible to detect faults. Differential relays perform the comparison on each phase. In the following description, only one phase of a three phase system is considered, but a practical relay contains three separate channels to evaluate faults on each phase or between phases.

In a numeric differential relay, the sampled current values need to be filtered and converted into a form suitable for comparison at the line ends. A simple and effective approach for achieving this is based on a Fourier technique. It will be recalled from Chapter 2, in Section 2.4, 'Spectral analysis', that the discrete Fourier transform (DFT) is a method for evaluating the frequencies contained within a series of N sampled data values at a frequency resolution of $1/NT$ where T is the sampling interval and N the number of samples. Unlike the fast Fourier transform, it is possible with the DFT to evaluate only one frequency of interest; for example, for a sampling rate of 400 Hz, if $N=8$ and $m=1$ in Equation 2.6 of Chapter 2, then $X(1)$ will represent the 50 Hz content of the sampled signal. Note also that $X(1)$ is a complex quantity and contains a real and imaginary component of the 50 Hz content. This thus provides a basis for both filtering and converting the current into a form suitable for a Kirchoff current summation.

For a sampling frequency of 400 Hz, typical for a numeric current differential relay, the equations used to evaluate the real and imaginary components of the sampled current waveform are:

$$I_S = \frac{2}{N} \sum_{n}^{N-1} i(n) \sin [nT\omega] \tag{4.16}$$

$$I_C = \frac{2}{N} \left(i(0) + i(N) + \sum_{n}^{N-1} i(n) \cos [nT\omega] \right) \tag{4.17}$$

where I_S = sine or imaginary component of the current samples

I_C = cosine or real component of the current samples

ω = power system frequency = 50 Hz

$i(n)$ = sampled current value at time n

N = number of samples per power system frequency cycle = 8.

Equations 4.16 and 4.17 are derived from Equation 2.6, but there are small differences that enhance the frequency responses. The group delay associated with the formation of I_S and I_C is one cycle of power system frequency and, thus, a numeric current differential relay of this type does not give ultra-high-speed operating times.

4.2.4.2 Communication channel propagation time delay measurement

It will be seen necessary in the next section to find the propagation delay of the communication channel used to connect differential relays. In general, numeric differential relays are designed to work on 64 kbit/s digital communications channels (56 kbs in North America) or at a lower data rate if an analogue communications medium is used, such as a voice frequency link. In either case it is important that compensation is made for the propagation delay of the communication channel. It is possible for a numeric differential relay to measure its channel propagation delay continuously. The situation is shown in Figure 4.13 where a line is being protected by two digital differential relays situated at the line ends. The relays send each other their complex current values calculated from the previously described Fourier technique. Each relay sends and receives a packet of information, commonly 20 bytes long, at each sample interval. The 20 byte message will contain the current vectors, error checking information and timing data. Since the processor clocks of the relays at each end are not synchronised, and nor are they necessarily running at precisely the same frequency, it is necessary to ensure that the current vectors from each end of the line are compared at the same point on wave. This is achieved as follows.

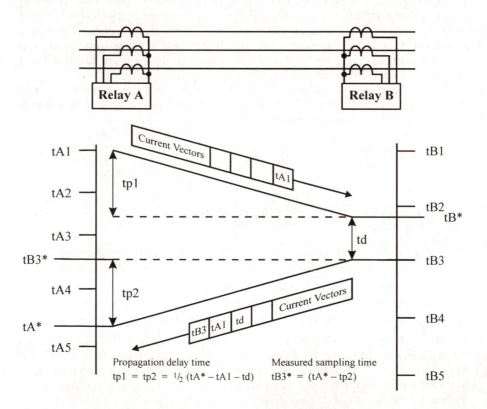

Figure 4.13 Communication between current differential relays

Relay A makes its samples at times t_{A1}, t_{A2} etc, and relay B similarly samples at t_{B1}, t_{B2} etc. Note that the sampling instants are different. At time t_{A1}, relay A sends a data message to relay B, contained in the message is time t_{A1} from relay A. Upon receipt of this message, relay B records the time of arrival as t_{B*}. Relay B then waits for time t_d until its next sampling instant occurs and then sends a message back to relay A. Included in the message to A is t_{B3}, the time at which B sends the message, t_d, the time from receiving a message from A to sending a message back, and t_{A1}, the time information that A originally sent to B. B's message to A is received by A at time t_{A*}. Since A knows when it sent the message to B, and knows when a reply was received, A can now calculate the propagation delay, t_p, assuming the delay is the same in each direction, from:

$$t_p = \frac{1}{2}(t_{A*} - t_{A1} - t_d) \tag{4.18}$$

In a similar fashion, A can calculate the time difference existing between its sampling instants and those of B. A may calculate the time, t_{B3*}, at which the last sample was made by B:

$$t_{B3} = (t_{A*} - t_p) \tag{4.19}$$

where t_{A*} is the time of arrival of B's message. Thus, A knows the time difference between the its most recent sample made at time t_{A4} and the nearest sample made by B. This time difference is important since the real and imaginary values of the measured current are time-varying and it is necessary to compensate for the difference to ensure correct relay behaviour. Note that the measurement of both the channel delay and sampling instant differences is made every time a message is received.

Automatic assessment of the propagation delay is suitable in situations where the communication media properties vary with time. This is possible in digital communications networks where the transmission path can be automatically rerouted if a network fault has occurred. If a dedicated communication channel, i.e. fixed propagation delay, is used with a numeric current differential relay, it is sufficient to program the relay with a measurement of the propagation delay made during relay commissioning.

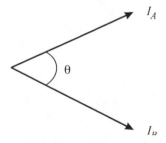

Figure 4.14 Phase difference between current values measured at ends A and B of a line

4.2.4.3 Time alignment of current vectors

Figure 4.14 shows the current vectors I_A, taken at time t_{A4} relay A, and I_B, taken at time t_{B3*} relay B, drawn in a phasor diagram. The angle θ between them is given by:

$$\theta = 2\pi f(t_{A4} - t_{B3*}) = 2\pi f(t_{A4} - t_{A*} - t_p) \tag{4.20}$$

where f is the power system frequency. To make these vectors cophasal, it is necessary to advance I_B forward in time until it is coincident with I_A. Thus, the time-advanced value of I_B, I_{Badv}, becomes:

$$I_{Badv} = I_B\, e^{j\theta} = I_B\,(\cos\theta + j\sin\theta) \tag{4.21}$$

Note that since I_B exists as a vector of real and imaginary components, this calculation is relatively simple. The values of $\cos\theta$ and $\sin\theta$ will be stored in a look-up ROM to avoid the need for recalculating the coefficients every time.

In a practical relay, the current vector I_B is also moved backwards until it is coincident with the previous I_A current vector. Thus, the required communications bandwidth is halved because each current vector can be used twice.

4.2.4.4 Relay characteristic

The relay calculates differential and bias current values in a similar fashion to a conventional current differential relay. For a two-ended line with ends R and S, the equations are, for say the 'a' phase:

$$|\,I_{diff}\,| = |\,I_{Ra} + I_{Sa}\,| \tag{4.22}$$

$$|\,I_{bias}\,| = \frac{1}{2}\left(|\,I_{Ra}\,| + |\,I_{Sa}\,|\right) \tag{4.23}$$

The absolute value $|\,I\,|$ of current vector I is given by

$$|\,I\,| = \sqrt{I_s^2 + I_c^2} \tag{4.24}$$

The evaluation of a square root is time-demanding task for a microprocessor to perform and is thus better avoided. In place of a square root, one approach is to use a least squares approximation technique or, alternatively, since the absolute values are required, the fault evaluation can be performed using the squares of the signals.

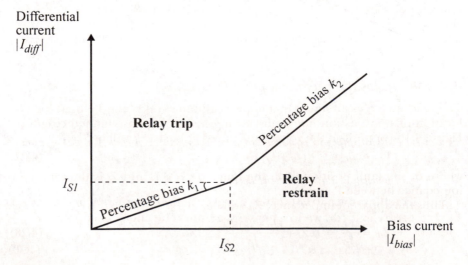

Figure 4.15 Biased differential characteristic

Thus, a standard percentage bias characteristic may be implemented as shown in Figure 4.15 where the lower percentage bias tripping criterion is:

$$|I_{diff}| > k_1 |I_{bias}| + I_{s1} \tag{4.25}$$

where k_1 is the percentage bias setting and I_{s1} is the minimum differential current setting and the higher bias criterion, where $|I_{bias}| > I_{s2}$, is:

$$|I_{diff}| > k_2 |I_{bias}| - (k_2 - k_1) I_{s2} + I_{s2} \tag{4.26}$$

Different implementations of numeric differential relays use different sampling frequencies. However, if a low sampling frequency is used, e.g. 400 Hz, then the requirements for a counting strategy, that is deciding how many consecutive samples of indicated fault should cause the relay to trip, are less demanding than in the cases of the numeric distance and directional comparison relays where sampling frequencies in the kHz range needed careful consideration. As with all differential relays, numeric implementations have no independent operating mode and so are entirely reliant on the digital communications path between relays.

4.3 Fault location

4.3.1 Introduction

The application of numeric techniques to protection relays has also led to the development of numeric fault-locating devices. This development is timely since increasing commercial and environmental pressures are forcing power utilities to drive their networks harder with the result that permanent faults must be dealt with in the shortest possible time.

Numeric techniques are ideally suited to fault location. Taking the case of a distance relay, implementations of this relay prior to numeric types could only indicate whether or not a fault lay within the characteristic. A numeric relay, however, is able to calculate the apparent impedance of a fault. Since this impedance is related to the distance, it is therefore useful as a fault-locating aid. Although it will be seen that this technique is not particularly accurate under all conditions, the flexibility of the numeric approach allows better algorithms to be implemented and thus increases location accuracy.

4.3.2 Fault location using apparent reactance

The reactance measurement of a numeric distance relay may be used for the location of transmission line faults. Using the single phase faulted circuit of Figure 4.16 (opposite), the voltage measured at busbar P will be:

$$V_p = I_p \alpha Z_l + (I_p + I_q) R_f \tag{4.27}$$

where α, the fault position, is in the range $0 < \alpha < 1$, and Z_l is the total line impedance between P and Q.

Thus the apparent impedance, Z_a, will be:

$$Z_a = \frac{V_p}{I_p} = \alpha Z_l + \frac{(I_p + I_q) R_f}{I_p} \tag{4.28}$$

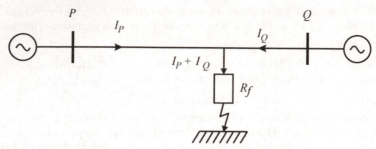

Figure 4.16 Faulted single phase transmission line

which is seen to depend on the fault current measured at end Q, i.e. the remote end infeed. If I_p and I_q are in phase, then the quantity $(I_p{+}I_q).R_f/I_p$ will be entirely resistive and the fault position α may be estimated from:

$$\alpha = \frac{X_a}{X_l} = \frac{\text{imaginary part of } Z_a}{\text{imaginary part of } Z_l} \qquad (4.29)$$

In practice the busbar fault currents will only be in phase if there is no pre-fault power flow. In the situation that a phase angle of δ exists between I_p and I_q, Z_a will appear as shown in Figure 4.17. This technique is seen to give accurate results only in cases of zero fault resistance or zero pre-fault load. As such it is of little practical use.

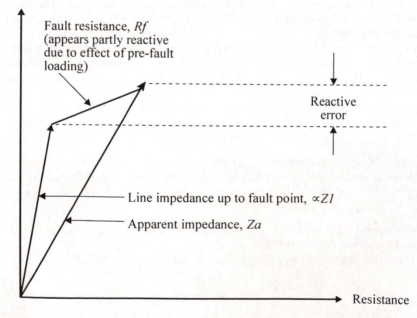

Figure 4.17 Effect of fault resistance and pre-fault loading on apparent impedance

4.3.3 Compensation for remote end infeed

It is possible to compensate for the remote end infeed [5] by using the superimposed component of the measured current at end P. Equation 4.27 is rewritten as:

$$V_p = I_p \, \alpha \, Z_l + \frac{I_{pf} R_f}{Da} \tag{4.30}$$

where $I_{pf} = I_p - I_{load}$, the superimposed fault current at P, and I_{load} is the pre-fault load current, and Da is an expression referred to as the *current distribution factor*.

$$Da = \frac{I_{pf}}{I_f} = (1 - \alpha) \, Z_l + \frac{Z_{sq}}{Z_{sp} + Z_l + Z_{sq}} \tag{4.31}$$

where Z_{sp} and Z_{sq} are the source impedances at ends P and Q of the line, respectively. Combining equation 4.30 and 4.31 gives:

$$\alpha^2 - \alpha \, k_1 + k_2 - k_3 R_f = 0 \tag{4.32}$$

where

$$k_1 = \frac{V_p}{I_p Z_l} + 1 + \frac{Z_{sq}}{Z_l} \tag{4.33}$$

$$k_2 = \left(\frac{V_p}{I_p Z_l}\right)\left(1 + \frac{Z_{sq}}{Z_l}\right) \tag{4.34}$$

$$k_3 = \left(\frac{I_{pf}}{I_p Z_l}\right)\left(1 + \frac{Z_{sq} + Z_{sp}}{Z_l}\right) \tag{4.35}$$

Thus, by equating the imaginary terms in equation 4.32, an estimate for α may be found. This approach may be readily extended to include a three-phase power system representation based upon symmetrical component theory. The algorithm suffers from two drawbacks:

(a) A value for the remote end source impedance, Z_{sq}, is required. This impedance can rarely be specified with accuracy and is liable to change with time. The only solution to this problem is to use a two-ended measurement approach where voltages and currents are recorded during the faulted period at both ends of the line.

(b) The algorithm neglects the effect of shunt capacitance. Thus its accuracy on long lines where the capacitance is significant will be degraded.

Other approaches have been proposed [6] where a compensation for shunt capacitance is made following an estimate of the fault position calculated using a conventional analysis similar to the foregoing. However, due to the distributed nature of capacitance along a transmission line, accurate results will only be forthcoming if the power system model used in formulating the solution contains shunt capacitance.

4.3.4   *Accurate compensation for shunt capacitance*

The most accurate fault location method uses two-ended measurements and the hyperbolic representation for a transmission line [7]. The hyperbolic pre-fault representation of line PQ in Figure 4.16 may be stated as:

$$\begin{bmatrix} V_p \\ I_p \end{bmatrix} = \begin{bmatrix} A & B \\ C & D \end{bmatrix} \begin{bmatrix} V_q \\ I_p \end{bmatrix} \tag{4.36}$$

By manipulation of equation 4.36 using information from ends P and Q of the line, taken both before and during the fault, an expression for the fault position may be derived. This expression is necessarily complicated, especially when the three-phase nature of the line is included, and so the mathematics will not be described further. This method is the most accurate approach to fault location since it uses two-ended measurements and the most realistic transmission line model. The result it produces is as accurate as the specification of the line parameters.

In practice, this theory has to be expanded to include three phases and all possible fault types. The theory for achieving this is based upon the work of Wedepohl [8] and is outside the scope of this text.

4.3.5   *Hardware for fault locators – fault recorders*

There are three possible forms for the hardware of a fault locator to assume. These are:

(a)   As a stand-alone device – the fault locator has its own dedicated hardware similar in design to numeric relay hardware;
(b)   As an adjunct to a numeric relay – in this case, the fault locator algorithm is included together with the relay protection algorithm and executed, on command, by the relay processor;
(c)   In conjunction with a digital fault disturbance recorder – this approach appears to be favoured commercially and will be considered further.

Fault recorders are used for a variety of reasons which include:

(i)   Monitoring protection operating time;
(ii)   Monitoring circuit breaker operating time;
(iii)   Monitoring circuit breaker currents;
(iv)   Fault location.

A fault recorder makes a permanent record of power system quantities such as voltages, currents and circuit breaker states. Currently, fault recorders are implemented digitally. Being microprocessor-controlled, it is possible for a digital fault recorder to be programmed to make automatic assessments of the uses given above. However, this tends not to be the case and, instead, it is more common for the fault recorder to send its recorded information to a remote location where it is analysed by a personal computer. This transfer of information is commonly performed by a serial data transmission over a voice frequency telephone line. It is becoming standard practice for utilities to install fault recorders, fitted with modems, at key substations. The fault recorders are permanently connected to telephone lines which enables their records to accessed by 'dialling up' the fault recorder and communicating through the use of a personal computer running appropriate software. If fault location is based upon this system, then the fault location software is targeted to run on a personal computer rather than the fault recorder. This eases the processing requirements on the fault recorder since there is no requirement to perform any filtering or conversion of sample values into phasor form.

The hardware of the fault recorder is similar to the hardware of a numeric relay as shown in Figure 4.1 and described in Section 4.1. However, there are several differences:

(1) Owing to the lower processing requirement, a fault recorder will usually have only one microprocessor;

(2) In addition to recording analogue information, digital information, e.g. the state of a protection trip output, is also recorded. Thus, a fault recorder will also include some form of digital input to this effect. Another purpose of this form of input is to trigger the recorder under certain conditions of interest. Thus, there is some form of programming of the trigger inputs, by the user, to achieve the desired response;

(3) The RAM of the fault recorder will be much larger than that of a digital relay to enable storage of the relevant records;

(4) A fault recorder may also have some form of mass storage device, such as a hard disk, to which the records are transferred when the RAM is full. Note that it is common for the fault recorder sampling frequency to be programmable by the user. The sampling frequency chosen will influence the RAM and disk storage limitations.

4.3.6 Phasor extraction

The algorithms described for fault location all use phasor values of the line voltages and currents. The simplest method of forming phasor values from sampled data values is to use the Discrete Fourier Transform evaluated at power system frequency, as described in Chapter 3. The result from the DFT will be in terms of real and imaginary parts which can then be readily converted to phasor form where necessary.

Unlike a digital relay where the digital filtering function is compromised with the relay operating time, in fault location there is little restriction on filter group delays. Thus, in addition to the phasor extraction, extensive digital filtering of the sampled values prior to the application of the DFT may be applied to ensure removal of all non-50 Hz components.

4.4 Software considerations

Much emphasis has been placed within these pages on the hardware of numeric relays. However, an increasingly large part of a numeric relay development project is spent on software development and so it is appropriate to describe the stages involved in this process.

On reception of a software specification for inclusion in the design of a relay, the software designer carries out a detailed analysis of *what* is required in the specification. There must be full understanding of the specification's intent and the designer must have the foresight to ensure that the requirements can be met physically before starting on the software design (see Figure 4.18).

Once this is done, the designer uses computer-aided software engineering (CASE) tools such as 'Select' to create a structured software design. This enables a neater design which is easily modified should the need arise and promotes good documentation throughout the development which is essential to any software project. Having produced a design which meets the specification's requirements and has the approval of the other team members, such as the hardware designer, the equipment on which the software is to be implemented in the relay is examined. An implementation model for the software is then developed using CASE tools.

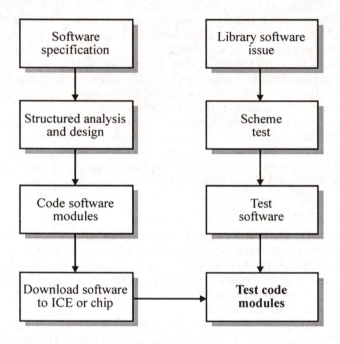

Figure 4.18 Software development

The software design is taken and coded in modular form. Each module should be small in length and well-documented so it is easy to understand. It is very important to comment the code in a detailed fashion. Thus, if any modifications need to be done at a later date, it will be easier to understand the current program and to implement the change. Coding may be in languages such as assembler, C or others, depending on the development tools available. Modular coding also makes testing the software much easier. Each module is tested for errors and, initially, much of this testing can be carried out on the personal computer (PC) on which the code has first been written. Every time a module is changed the original design needs to be updated, and by virtue of its structure, a good software design will highlight other areas of the design which may also need changing as a result. Each version of the software is stored on the mainframe by a means such as the software management system (SMS). A new version of the software is created each time a change is made and this is put away on the SMS under the appropriate version number. Each modified module must consequently be re-tested for errors.

Development tools such as compilers, assemblers, linkers, locaters and in-circuit emulators greatly aid the process of transferring the coded software from the PC to the chip(s) on which it is to run in the hardware. Different chips require different tools. For instance, the 8051 8-bit microcontroller needs an 8051-specific compiler, assembler, linker and locater which will produce an object or hexadecimal code from the C code and/or the assembler code which can be understood by the chip. Furthermore, this hexadecimal code will be arranged by the linker and locater such that the chip carries out the program's instructions in the intended order.

Depending on the quality of the original software design, a varying amount of debugging will be required to bring the software as a whole to an errorless state. An in-circuit emulator (ICE) is invaluable for this purpose. The program can be downloaded into the ICE which is connected to both the relay and a PC and can then be stopped at various places, or registers may be examined to see if they contain the expected information. If an ICE is not available for the chip in question, the code will need to be downloaded to the chip itself which is placed in the relay, and a logic analyser used to probe the pins of the chip for data flow. Again, if bugs are found in the software, the original design must be updated and brought into line, along with the code, and the software re-tested as a whole.

Different chips are used as memories for the numeric relay's information store. The software itself is normally downloaded into an EPROM chip by means of a PROM programmer such as the 'Unipak 28'. RAM is used for holding current data such as the status of flags, or the value of a variable. Another memory type is the E^2PROM which is used to store the values of settings needed on first initialising the software. The E^2PROM holds the value of the registers indefinitely but these register contents can be changed as well as read, unlike the EPROM which is a read-only memory type. These memory types may be accessed by the microprocessor chip if the hardware scheme enables this.

When the designer is satisfied that the software works as required, it is time to carry out scheme tests on the equipment as an entity. The relay must be seen to perform according to its overall specification, to function correctly and within the timescales required. Any rework at this stage has implications for the software and this must be re-tested with each modification. The importance of documentation at each stage of the development cannot be stressed enough, in order to produce a controlled product which meets its specification.

The final step for the software is for it to be issued to the production library files where it can be accessed by the production department and downloaded in its final form to the appropriate chip which is placed in the relay. Further software development at this point entails retrieving the files from the library with suitable notification to relevant personnel and reissue of the software under a different version number when the new development is complete.

4.5 Numeric relay testing

4.5.1 Introduction

The advent of digital technology has not only produced a range of advanced protection relays but has also allowed the utilisation of more realistic relay testing facilities. It is especially pertinent that numeric relays, which generally give superior performance under distorted waveform conditions than more conventional relays, are tested under realistic conditions. The advantages of computer-based relay test sets include greater bandwidth of faulted relaying signals, versatility of power system structure and automated testing [9].

4.5.2 Relay test hardware

The general outline of a computer-based relay test set is shown in Figure 4.19. The set acts, in essence, like a compact disc player. Previously calculated power system fault data are sent to digital to analogue converters which are connected to suitable voltage and current amplifiers capable of providing realistic relay input levels.

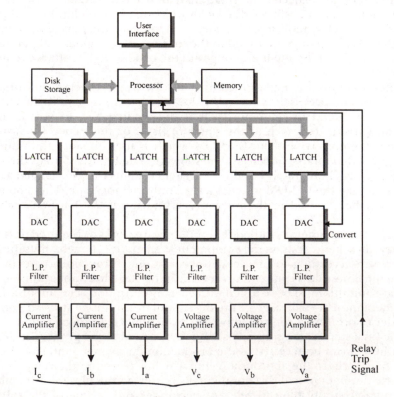

Figure 4.19 Computer-based relay test hardware

Time series simulation data are stored on the disk of the system. Very accurate fault simulation requires complex software which would ordinarily be run on a mainframe computer and then transferred to the test set using floppy disk. The task of the test set processor is to transfer the data from the disk to the DACs at the required sampling frequency. Since sampling frequencies can be as high as 12 kHz, it is usual for the simulated data to be transferred first into the processor memory (RAM) and then into the DACs to overcome the slow access speed limitation of disk drives. The word length of the DACs must be at least the same, but preferably greater than, the word length of the ADCs used in the relay under test. The DACs found in such relay test sets have 16 bit word lengths. Another important task for the processor is to convert the data samples, which will normally be expressed as floating point numbers, into 16 bit values suitable for the DACs.

Each of the DACs is preceded by a latch, i.e. simple one word storage. This is to ensure that all the DAC outputs change from one sample to the next at exactly the same instant. The processor loads each latch with the relevant data and then instructs all the DACs to convert simultaneously using the common CONVERT line. A low-pass filter is placed after the DACs to remove the steps in the analogue waveforms caused by the change from one sample to the next. This filter essentially performs the same function as the anti-aliasing filter used in a numeric relay input stage as described in Section 4.1.1; the filter has a passband from d.c. to half that of the output sampling frequency.

The final stage of the test set is the output amplifiers. The voltage amplifiers give an output of typically 110V. The current amplifiers, which are strictly speaking transconductance amplifiers since the input from the DAC is a voltage and the amplifier output is a current, will be able to give peak current levels of 100A. The test set shown in Figure 4.19 has 6 output channels, 3 for voltage and 3 for current; it is thus capable of testing distance protection. However, more channels and different amplifiers can be added to enable testing of any relay. For example, changing the 3 voltage amplifiers to current amplifiers will allow two-ended differential scheme testing.

In addition to the output signals to the relay, an input signal to the test set to record the relay trip contacts is also included. This allows the test set to monitor the relay operating time accurately.

A user interface is provided in the test set and will usually take the form of a visual display unit and keyboard. This relay test equipment is also capable of playing back data acquired from a real system using a digital fault recorder. Although it was stated earlier that a mainframe computer would be used to calculate the simulated fault data, with current advances in personal computers (PCs), it will soon be possible for a PC not only to run the simulation software but also process the relay testing in addition.

4.5.3 Digital power system fault simulation

Conventional relay test sets use a series resistance and inductance to model a transmission line. Such a model will only yield 50 Hz and exponential offset information under fault conditions. In practice transmission lines are composed of series resistance and inductance and shunt capacitance infinitely distributed along the length of the transmission line. This highly accurate representation leads to the presence of travelling waves occurring on the system when a fault occurs at any point on wave other than close to a zero crossing. Travelling waves propagate outwards from the fault point and become reflected at discontinuities at the line ends.

When digital computers became commonplace, techniques were sought to be able to simulate the travelling wave effect. An initial approach was based upon a method called the Lattice Diagram, where all the waves are individually accounted for and the voltage and current waveforms at any point on the line are evaluated by superimposing all the travelling waves. The Lattice Diagram approach can be extended to include the effect of induced travelling waves on the sound phases, but it is not possible to model the attenuation of the waves effectively due to the series line resistance; the Lattice Diagram approach assumes a lossless line. A generalised approach to travelling wave solution was made by Wedepohl in 1963 [8] who developed the necessary theory. A time-stepped solution to Wedepohl's theory was reported by Dommel and Scott-Meyer [10] who developed the Electro-Magnetic Transients (computer) Program, EMTP [11]. The EMTP has now become the standard program for simulating power system faults and EMTP output may be down-loaded into a numeric relay test set, as described in the previous section, for effective and realistic relay testing. In addition to transmission line fault simulation, the EMTP also allows accurate modelling of load flow, switching transients, generator behaviour and transient stability.

4.6 References

1   MOORE, P.J., and JOHNS, A.T.:'Distance protection of power systems using digital techniques', *IEEIE Electrotechnology*, Oct/Nov 1990, pp. 194–198
2   CROSSLEY, P.A. *et al.*:'The design of a directional comparison protection for EHV transmission lines', *IEE Conference Publication*, No. 302, 1989
3   LANZ, O.E. *et al.*: 'LR91 – an ultra-high-speed directional comparison relay for protection of high-voltage transmission lines', *Brown Boveri Review*, **1–85**, pp. 32–36
4   KWONG, W.S., *et al.*: 'A microprocessor-based current differential relay for use with digital communication systems', *IEE Conference Publication* No. 249
5   ERIKSSON, L., SAHA, M.M. and ROCKEFELLER, G.D.: 'An accurate fault locator with compensation for apparent reactance in the fault resistance resulting from remote end infeed', *IEEE Trans.* 1985, **PAS-104**, pp. 424–435
6   TAKAGI, T., YAMAKOSHI, Y., YAMAURA, M., KONDOW, R and MATSUSHIMA, T.: 'Developments of a new type of fault locator using the one-terminal voltage and current data', *IEEE Trans.* 1982, **PAS-101**, pp. 2892–2898
7   JOHNS, A.T., *et al.*: 'New accurate transmission line fault location equipment', *IEE Conference Publication* No. 302, 1989
8   WEDEPOHL, L.M.: 'Application of matrix methods to the solution of travelling-wave phenomena in poly-phase systems', *Proc. IEE*, 1963, **110**, (2), pp. 2200–2212
9   WILLIAMS, A. and WARREN, R.H.J.: 'Method of using data from computer simulations to test protection equipment', *IEE Proc.*, 1984, **131** (7), pp. 349–356
10   DOMMEL, H.W. and MEYER, W.S.: 'Computation of electromagnetic transients', *Proc. IEEE*, 1974, **62**, (7)
11   Electromagnetics transients program reference manual, H. Dommel, 1986

4.7 Further reading

'Tutorial course – microprocessor relays and protection systems', IEEE publication No. 88 EH 0269-1-PWR, 1988

PHADKE, A.G. and THORP, J.S.: 'Computer relaying for power systems' (Research Studies Press, 1988)

WRIGHT, A. and CHRISTOPOULOS, C.: 'Electrical power system protection', (Chapman and Hall, 1993)

JOHNS, A.T. and SALMAN, S.K.: 'Digital protection for power systems' (IEE power series No. 15, 1995)

4.8 Appendix: Typical numeric relay specifications

4.8.1 Electrical environment

DC supply voltage

The relay or its associated power supply unit for use in a 110 V (nominal) d.c. supply system is required to operate satisfactorily with a d.c. supply voltage range 87.5 V to 137.5 V and to withstand a maximum voltage of 143 V.

The relay or its associated power supply unit for use in a 48 V (nominal) d.c. supply system is required to operate satisfactorily with a d.c. supply voltage range 43 V to 60 V.

The static measuring relays and protection equipment shall meet the requirements of IEC 255-11 and their performance shall not be affected under the following conditions:

(a)   Interruption to the d.c. auxiliary supply of duration up to 10 ms;
(b)   AC component (ripple) in the d.c. auxiliary supply up to 5% of rated value.

4.8.2 Insulation

Dielectric

The relay shall meet the requirements of the dielectric tests in IEC 255–5. The test voltage shall be selected according to the rated insulation voltage of the relay from Series C of Table 1 of IEC 255-5. The rated insulation voltage of the relays connected to the current transformers of high impedance circulating current differential protection shall not be less than 1 kV. All other relays shall have a rated insulation voltage of not less than 250 V.

The relay open contacts shall withstand a voltage of 1 kV.

Impulse voltage

The relay shall meet the requirements of the impulse voltage tests in IEC 255-5 with a test voltage of 5 kV.

4.8.3 Electromagnetic compatibility

The following requirements are applied to static measuring relays and protection equipment. These requirements may be applied to some electromechanical relays which are of high speed or high sensitivity.

50 Hz interference

The relay shall meet the requirements of a power frequency interference test.

1 MHz burst disturbance

The relay shall meet the requirements of the test in IEC 255-22-1 with severity class III.

Electrostatic discharge

The relay shall meet the requirements of the test in IEC 255-22-2 with severity class III.

Radiated electromagnetic field disturbance

The relay shall meet the requirements of the test in IEC 255-22-3 with severity class III. The test shall be carried out by using Test Method A and sweeping through the entire frequency range 20 MHz to 1000 MHz.

Fast transient

The relay shall meet the requirements of the test in IEC 255-22-4 with severity class IV.

Chapter 5
Coordinated control
C. Öhlén

Introduction

The operation of a power system includes a large number of functions which, in the past, have been carried out by separate equipment and systems involving different personnel. The main functions are:

- Voltage and frequency control;
- Control and interlocking of switching devices;
- Protection of high voltage equipment;
- Status indication and event recording;
- Condition monitoring and alarms;
- Fault location and disturbance recording;
- Auto reclosing and automatic restoration;
- Synchrocheck and synchronising;
- Automatic sequential control;
- Load shedding and load management.

In addition to the above functions, statistical data are also collected but for different purposes, mainly for planning and scheduling of operation and maintenance but also to help making decisions regarding modifying and extending the network.

Planning of scheduling and remote control is handled in regional or nationwide control centres. Together these functions can be described as power system management, see Figure 5.1 (overleaf), and can be divided into the following systems:

SCS	Substation Control System
PCS	Plant Control System
SMS	Substation Monitoring System
SCADA	Supervisory Control and Data Acquisition System
EMS	Energy Management System
DMS	Distribution Management System

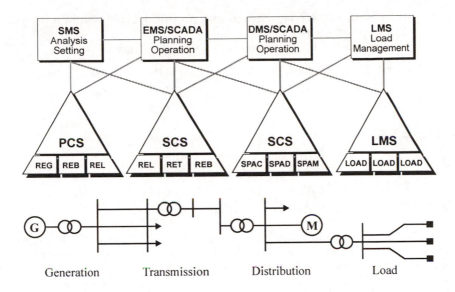

Figure 5.1   Power system management

The bay-oriented equipment for protection, monitoring, control and measuring of a transformer is designated *RET* and for a line is designated *REL*.

By coordinating these systems, design, operation and maintenance can be conducted more efficiently and economically than previously. It is, however, important to structure the systems so that basic requirements are not jeopardised; coordination is not the same as integration.

5.1 Conventional control systems

5.1.1 Functions and design

Conventional equipment for control, monitoring and protection in a substation has traditionally been designed with a combination of discrete static and electromechanical devices and includes a large number of separate subsystems. Control of switching devices, interlocking, alarms, disturbance recording, protection, measuring and metering are some of the subsystems and all of these require individual wiring. Designs of this type were originally used for round-the-clock manned substations. Even with the introduction of computers and SCADA this basic design structure has been used where the interface between the substation and the remote control centre is made via a remote terminal unit (RTU).

As an example of conventional control, the station depicted in Figure 5.2 will be discussed. For operational purposes the position of each breaker and isolator (disconnector) is monitored. When operating a breaker or a disconnector, either in the substation or remotely, it is necessary to check the position of all other switching devices. If, for example, an isolator is opened during heavy load, a severe fault may occur. An interlocking system is used to prevent this from occurring.

To open isolator I3A the following criteria must be fulfilled:

either (Figure 5.2a):	Operating switch	**Off**	and
	CB3	**Off**	and
	I3B	**Off**	
or (Figure 5.2b):	Operating switch	**Off**	and
	I2A	**On**	and
	I2B	**On**	and
	CB2	**On**	and
	I3B	**On**	

In a conventional station this logic is implemented with auxiliary relays and hardwiring. For the station of Figure 5.2 there are only 4 circuit breakers; it is easy to see that the logic will become complicated for larger stations. If other conditions such as voltage, synchrocheck and breaker status have to be included in the logic this will require an even larger number of auxiliary relays and contacts. This is also the case for sequential control such as load transfer or automatic restoration after a fault.

At each busbar the following quantities need to be monitored: voltage, current, frequency, real and reactive power. In a transformer oil level, oil pressure and temperature have to be monitored together with load data. Furthermore, many transformers have *on-load tapchangers* which need to be both controlled and monitored.

5.1.2 Disadvantages of using traditional technology

Conventional technology for substation monitoring and control has been described in the previous section. For unmanned or relatively large substations the use of 'conventional' technology will result in a large number of components such as auxiliary relays and many miles of wires. In general, this technology is costly and both time- and space-consuming which is detrimental to overall security and dependability since these factors decrease with the number of wires and components. The main disadvantage of conventional technology comes, however, during maintenance or whenever modification or extension of the control system is required. If a part of the system is taken out of service for maintenance, it may affect other parts of the system unless careful procedures are adopted. Fault tracing can cause long interruptions due to the large number of cables present – especially where cables are not marked correctly. The main disadvantages are summarised below:

- *Quality assurance:* In conventional control system design a large number of components and wires are needed to realise both simple and more complex functions. This has to be done individually for each substation and so standard and proven solutions cannot be used. Furthermore there is little redundancy in such designs and so a single component failure can affect the whole system giving a poor degradation withstand capability.

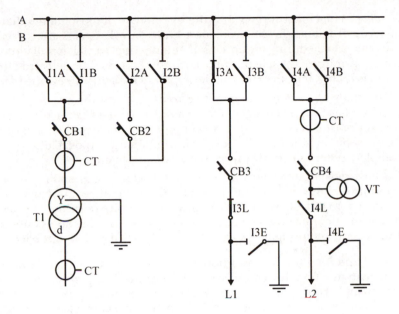

Figure 5.2a Interlocking requirement to open I3A with circuit L1 on load (on load busbar changeover)

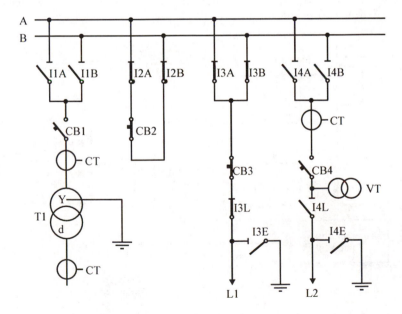

Figure 5.2b Interlocking requirement to open I3A with circuit L1 on load (on load busbar changeover)

- *Installation and commissioning:* It is the experience of most utilities that the majority of failures in the protection and control systems are caused by human mistakes during the process of engineering, wiring, installation and commissioning of the system. Some of these errors may be detected during commissioning, but others may not be apparent until they have caused a fault.

- *Maintenance and testing:* Maintenance to any part of the system normally requires hardwiring changes. For example, to maintain a breaker, the breaker must be taken out of any event-logging and alarm systems as well as any sequential logic. It is, furthermore, almost impossible to test the complete system since it involves so many components.

- *Expandability:* Modifications in an existing station, e.g. adding an extra transmission line, are difficult since a large part of the control logic in the station has to be changed which can involve extensive rewiring. This, again, runs the risk of new errors being introduced besides being time-consuming and possibly requiring that part of the station is taken out of service.

Figures 5.3 and 5.4 show a conventional substation with separate subsystems for each main function. It is easy to see that the control and protection systems have many similarities where many functions are redundant. By coordinating these main functions it is possible to maintain separate control and protection functions but at the same time reduce the wiring, as depicted in Figure 5.5. This coordination can both decrease the required space and increase the overall reliability and availability of the control system. The remaining part of this chapter will discuss such a coordinated concept based on microprocessors and fibre optic communications.

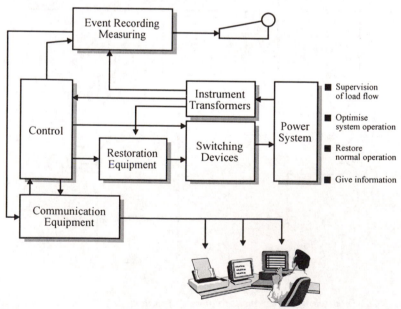

Figure 5.3   Requirements of a conventional control system

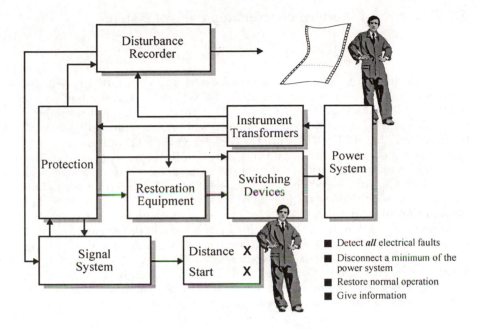

Figure 5.4   Requirements of a conventional protection system

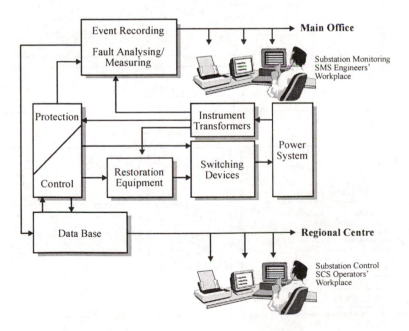

Figure 5.5   A coordinated protection and control system

5.2 Concepts of modern coordinated control systems

5.2.1 System architecture (distributed processing)

Figure 5.6 shows an example of a coordinated and distributed system for control and protection encompassing the SCADA/EMS and the bay. In the control centre a local area network (LAN) is used for communication. For redundancy the LAN can be duplicated. A typical LAN for this application will be based on the IEEE 802.2 and IEEE 802.3 (Ethernet) standards giving synchronous communication running at 10 Mbps on a coaxial cable. Communication to the various substations can be achieved using a wide area network (WAN) running at, for example, 4800 bps. The different substations can include older systems with RTU and newer systems with SCS. For transmission and subtransmission, a two level structure with *station bus* and *object bus* is used. The station bus uses the same LAN as used in the Regional Centre. It is not usually duplicated (although it could easily be) because of the inherent high availability of the system. Due to the high disturbance level found in the station, an active fibre optic star is used. On the limited object bus, which normally has a maximum length of 10 m, an optical ring is used with asynchronous communication running at 9600 bps.

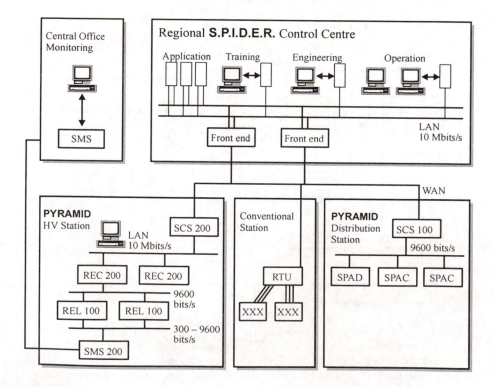

Figure 5.6   ABB network control and protection

For transmission of disturbance recorder information through the SMS, a separate fibre optic ring is used as a *monitoring bus* which is directly programmable to 300-9600 bps and so can be matched with the speed of the telephone system. This allows a commercially available telephone modem to be used for communication to the main office. For a distribution station a centralised solution using only the 9600 bps fibre optic ring can be used. For larger distribution system the two-level concept with a station bus may be more suitable.

Figure 5.7 shows an example of a small substation with two incoming transmission lines, one transformer and two feeders. With distributed processing at the station level, no central station computer is required. The LAN provides a democratic system with no master-slave relation. Information is exchanged between the various bay control terminals (REC) continuously. Both a gateway (i.e. access to a commercial data transmission service) and a workstation can be connected independently to the LAN. If, for example, the gateway connection is broken the station can still be operated from the local workstation. It is also possible by this configuration to access all REC terminals from any of the REC terminals.

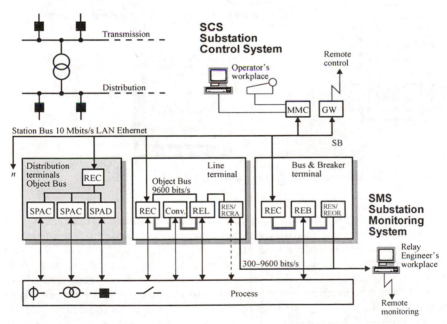

Figure 5.7   The intelligent substation

The control, protection, monitoring and measuring functions can be structured according to Figure 5.8. The bay control terminal, REC, handles all communication between the bay level and the substation level where events are also time-tagged. It also includes all the necessary software functions for the control. The control terminal has several I/O possibilities including both transducer inputs and I/O for digital signals to hardwired contacts. Each REC has four numeric inputs which can run two different protocols simultaneously. This allows flexibility to connect both conventional and numeric devices as well as devices from different manufacturers.

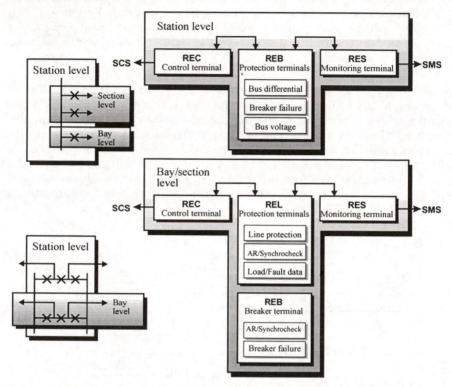

Figure 5.8   Modular structure of a coordinated system

In a completely numeric SCS, including numeric protection, all communications at bay level will take place on the object bus or the monitoring bus. For example, the line terminal REL will provide numeric data for line current, voltage, frequency and both real and reactive powers. Furthermore all indications, as well as fault data, fault location, voltages, currents and phase angles, will be provided after a disturbance. The disturbance recording function will be available through the RES monitoring terminal on the SMS in order not to load the SCS unnecessarily with a large amount of data. Furthermore the SMS function can be seen as a back-up to the SCS in the case of a severe disturbance.

For a single breaker switchgear, each line typically has one REC terminal and a duplicated REL line protection terminal. The disturbance recording function RES is not normally duplicated. For the breaker, a separate breaker terminal REB can be provided. Some functions can be provided within different terminals depending on user preference. For example, autoreclosing can be included either in the REC, REL or REB terminal. The breaker failure function can also be located in different terminals including centralised solutions together with the bus protection. The use of this modularised and distributed structure ensures a very high degradation withstand capability. If one function fails it will not affect others. If, for example, an RES terminal fails, it is still possible to obtain disturbance data from the other bays.

For a $1\frac{1}{2}$ breaker station the bay control terminal REC would communicate with up to 4 RELs, 2 RESs and 3 REBs. If more devices need to be connected to the object bus, then this is possible within limits.

5.2.2 Numeric technology

With numeric technology all control functions are available as software modules. The software in the REC control terminal is divided into two main groups: *system software* and *application software*.

The system software contains modules that determine the basic functions of the system. This software is placed in a non-volatile memory and cannot be changed by the application engineer. The following are included in the system software:

- *Real-time operating system* which contains task management, time management, system start-up and error handling;
- *Basic input/output system* which provides a device-independent interface to peripheral units such as printers, keyboards and VDUs;
- *Console communication subsystem* provides tools for test and maintenance of the computer system;
- *Process communication software* contains functions that handle the input/output boards connected to the process as well as time-tagging at status change or limit passage included in the process;
- *Process control environment software* contains functions for creating and executing control functions.

The application software adapts the basic SCS to a specific station configuration or the user's requirement for specific functions. The main part of the application software is standard function modules, each module containing a well documented and tested function which could be either a specific control or supervising function. The remainder of the software contains the interconnection between standard modules or user-specific functions. The user programming of the SCS should be done in a simple high-level process language with graphic representation. The function of a block can be simple, such as an OR function, or more complex such as a sequence (for example bus transfer)

The following functions can be included as standard modules in the SCS software:
- Apparatus control, e.g. one breaker;
- Bay control, e.g. a sequence for a complete bay;
- Interlocking;
- Sequences, e.g. bus transfer;
- Tapchanger control;
- Parallel control of transformers;
- Automatic reclosing.

5.3 System functionality

5.3.1 Bay level

At bay level the bay control terminal is used to coordinate all functions within the bay. In $1\frac{1}{2}$ breaker stations the bay control terminal coordinates the whole function. The bay control terminal contains functions that can be placed in common hardware without violating the demands on security, dependability and degradation withstand capability. Such functions include:

- Data acquisition;
- Control of HV apparatus;
- Interlocking of HV apparatus;
- Bay level sequences;
- Automatic reclosing;
- Voltage control performed by tapchangers, reactors or capacitor banks.

At bay level the protection equipment is separate but can communicate with the REC terminal.

5.3.2 Substation level

The workstation for man-machine communication (MMC), together with the gateway, is normally the only equipment connected at the substation level. The computerised MMC replaces the conventional control panel as well as other functions such as the alarm panel and event recorder. A number of displays are available to inform the operator of the status of the process and the SCS. The following types can be available.

- Process displays (e.g. single line diagrams);
- Report displays;
- SCS maintenance displays;
- Alarm and events lists.

At substation level, from the MMC, the following functions can be carried out by the operator:

- Operation of high voltage apparatus;
- Indication of status;
- Indication of measured values and limit check;
- Operation of automatic control equipment;
- Fault indication;
- Sequential events handling;
- Parameter reading and setting of protective relays.

If required a separate substation control terminal can be connected to the station bus. This could handle the following functions:

- Automatic restoration (of the complete substation);
- Report calculation;
- Centralised sequence operation;
- Supervision and control of auxiliary station equipment.

5.4 Man-machine interfaces (MMIs)

5.4.1 Bay level

Bay level control only provides an indication of the status of the equipment. It is, however, possible to install a redundant back-up panel which can operate breakers and disconnectors at the bay level. In such cases station interlocking is still provided from the SCS. The protection terminal is normally provided with a built-in display which can be used to read indications as well as setting the protection.

5.4.2 Substation level

As described in Section 5.3, a workstation and/or a personal computer (PC) is used as a man-machine interface. On this workstation a number of different displays can be obtained as shown in Figures 5.9–5.15.

Selection of the different displays and menus can be made from the operator's keyboard, where it is also possible to execute different commands. To obtain the highest possible reliability, commands must be *selected* before being *executed*. When an object is selected, for example a circuit breaker, it is reserved in the SCS together with all other apparatus involved in the interlocking. Figures 5.16 and 5.17 show two types of keyboards used for these tasks. Finally, at the station level, a printer can be used for hard copies. Mass storage of data is normally not required at the station level since the station is connected to a regional control centre with this capability.

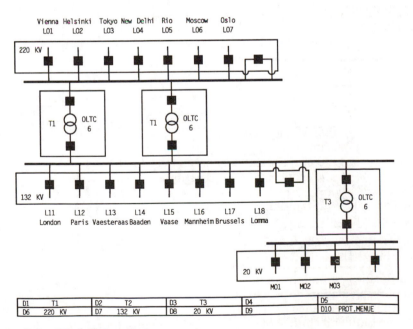

Figure 5.9   Workstation display: overview

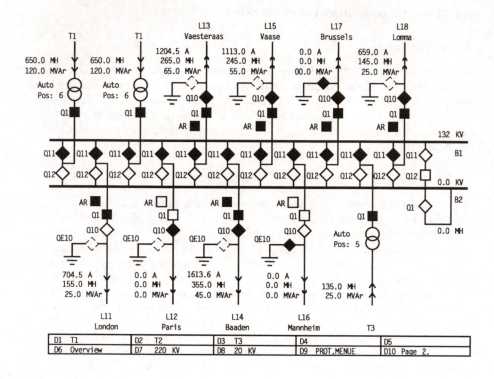

Figure 5.10   Workstation display: single line diagram

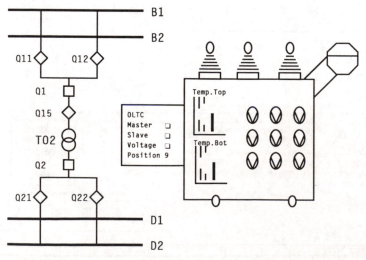

Figure 5.11   Workstation display: bay display

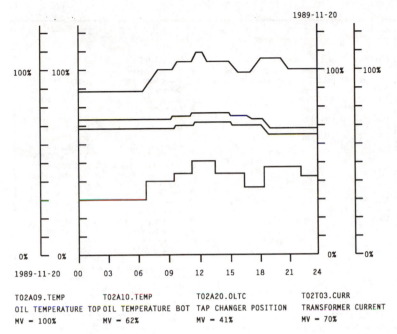

T02A09.TEMP	T02A10.TEMP	T02A20.OLTC	T02T03.CURR
OIL TEMPERATURE TOP	OIL TEMPERATURE BOT	TAP CHANGER POSITION	TRANSFORMER CURRENT
MV = 100%	MV = 62%	MV = 41%	MV = 70%

Figure 5.12 Workstation display: trend display

EVENT   PAGE 7(7)   1989-11-20   17:04:53

*	T02A11.GAS	BUCHHOLZ RELAY	Value	Alarm	15	13:46:16:542
	T02A11.GAS	BUCHHOLZ RELAY	Value	Normal	16	08:41:30:008
R	T02A11.GAS	BUCHHOLZ RELAY	Value	Alarm	16	08:41:34:340
R	T02A11.COOL	COOL FAN NOG	Value	Alarm	16	08:41:36:926
*	L13Q01	LINE BREAKER TO V-AS	Open	Order	17	10:11:26:145
*	L13Q01	LINE BREAKER TO V-AS	Close	Order	17	10:11:30:145
*	L13Q01	LINE BREAKER TO V-AS	Select	Order	17	10:11:32:745
*	L13Q01	LINE BREAKER TO V-AS	Remote	Order	17	10:11:34:845
*	L13Q01	LINE BREAKER TO V-AS	PowSwit	Alarm	17	10:12:15:445
	L13Q01	LINE BREAKER TO V-AS	PowSwit	Normal	17	10:12:17:945
*	L13Q01	LINE BREAKER TO V-AS	LowDensF	Alarm	17	10:12:18:445
	L13Q01	LINE BREAKER TO V-AS	LowDensF	Normal	17	10:12:20:345
*	L13Q01	LINE BREAKER TO V-AS	AuxSupp	Alarm	17	10:12:21:245
	L13Q01	LINE BREAKER TO V-AS	AuxSupp	Normal	17	10:12:23:345
*	L13Q01	LINE BREAKER TO V-AS	PowSwit	Alarm	17	10:12:24:345
	L13Q01	LINE BREAKER TO V-AS	PowSwit	Normal	17	10:12:25:945
*	L13Q01	LINE BREAKER TO V-AS	LowDensF	Alarm	17	10:12:26:545
	L13Q01	LINE BREAKER TO V-AS	LowDensF	Normal	17	10:12:27:345
*	L13Q01	LINE BREAKER TO V-AS	LowPresF	Alarm	17	10:12:27:945
	L13Q01	LINE BREAKER TO V-AS	LowPresF	Normal	17	10:12:29:145

D1 PRINT PAGE	D2 PRINT LIST	D3	D4 SECTION DISPL	D5
D6	D7	D8 ABORT PRINT.	D9	D10 BREAK

Figure 5.13 Workstation display: process event list

```
REL 100              DISTURB BUFFER  38    Date: 90-09-03   Time:14:19:50
                     UNIT NUMBER  31

PREFAULT             FAULT LOCATION        BINARY SIGNALS
V-L1  62.8 kV   6.7 Deg    Default fault loop  R    ■ Signal Autorecloser      Unsymmetry
V-L2  62.8 kV 247.0 Deg    Distance to fault    41%   Block Measuring          Synchro-check
V-L3  62.8 kV 127.0 Deg                              Carrier Signal            Loss of Voltage
V-E    0.0 kV 256.0 Deg                              Carrier Receive EF        Carrier Aided Trip
I-L1 502.5 A    8.8 Deg                            ■ Carrier Send            ■ Three Phase Trip
I-L2 501.9 A  248.4 Deg                              Carrier Send EF           Trip Earth Fault
I-L3 501.5 A  128.3 Deg                              DEF Forward             ■ Trip Phase L1
I-E    2.0 A  351.4 Deg                              DEF Reverse               Trip Phase L2
f     50.0 Hz                                        ECHO Weak Infeed          Trip Phase L3
                                                     EF Line Check             Trip Line Check
FAULT                                              ■ Single Phase Fault        Trip Stub Prot
                                                     Three Phase Fault         Trip Weak Infeed
V-L1   9.7 kV   6.5 Deg                            ■ General Start           ■ Trip Zone 1
V-L2  62.9 kV 247.0 Deg                            ■ Non Direct Start          Trip Zone 2
V-L3  62.8 kV 127.0 Deg                              Loss of Guard             Trip Zone 3
V-E    0.0 kV 251.6 Deg                              Power Swing Block         Trip Zone 3R
I-L1 999.2 A  288.2 Deg                            ■ Phase Selection E         Fuse Fail
I-L2   5.2 A  248.6 Deg                            ■ Phase Selection L1        Fuse Fail Trip
I-L3   3.2 A  127.5 Deg    COMMON                    Phase Selection L2        Self Supervision
I-E  995.80 A 287.9 Deg    □ Exit                    Phase Selection L3      ■ Zone 1
                                                     Start Non Dir EF        ■ Zone 2
                                                     System Supervision      ■ Zone 3
                                                     Overload                  Zone 3 Reverse


D1          D2                          D4          D5
D6          D7                          D9          D10
```

Figure 5.14 Workstation display: disturbance display

```
L1REL100                    REL100 Distance Protection

BASIC SETTINGS     UNIT NUMBER   31.00      ACTIVATE SAVED SETTINGS

LINE REFERENCE         CONFIGURATION         SETTINGS      □ BASE   ■ ALT
Station No.   0        □ CT Earth In          ZONE 1
Line No.      0        ■ CT Earth Out         Reactive Reach          10.00   10.00   Ohm
Unit No.      0        □ Indication Trip      Resistive Reach Ph-Ph   10.00   10.00   Ohm
Parall Line No. 0      □ Indication Start     Zero Seq Compensation   1       1.00
                       ■ Block Z1             Resistive Reach Ph-N    10.00   10.00   Ohm
TRANSFORMERS           ■ Block Z2             Time Delay              0       0       s
CT Primary   1000 A    ■ Block Z3             Block of Timer          □       □
CT Second.      0  A   □ Block AR
VT Primary    220  V                          ZONE 2
VT Second.    110  V   COMMUNICATION          Reactive Reach          20.00   20.00   Ohm
                       Scheme                 Resistive Reach Ph-Ph   20.00   0.00    Ohm
LINE                   □ Extension            Zero Seq Compensation   1       20.00
Frequency      50  Hz  ■ Zone 1 PUTT          Resistive Reach Ph-N    20.00   20.00   Ohm
Length        100      □ Zone 3 PUTT          Time Delay              300 0 s
Length unit     0      □ Zone 1 POTT          Block of Timer          □       □
Pos Seq X    12.0 Ohm  □ Zone 2 POTT
Pos Seq R     2.0 Ohm  □ Blocking             ZONE 3
Zero Seq X   48.0 Ohm  Unblock CRG            Reactive Reach          30.00   0.00    Ohm
Zero Seq R    8.0 Ohm  ■ No CMD Unblock       Resistive Reach         30.00   0.00    Ohm
Pos Seq X Loc Source 12.0 Ohm □ CMD Unblk Without Win  Zero Seq Compensation  1.00  0.00
Pox Seq R Loc Source  2.0 Ohm □ CMD Unblk With Wind.   Resistive Reach Ph-N  30.00  0.00  Ohm
Pos Seq X Rem Source 12.0 Ohm Unblock Time TO 20 ms    Time Delay            800    0    s
Pos Seq X Rem Source  2.0 Ohm                          Time Delay Reverse    2000   0    s
Mutual X Par Line     0.0 Ohm                          Block of Timer         □      □
Mutual R Par Line     0.0 Ohm SETTING MODE             Block of Timer Reverse □      □
                              Local
                              Remote

D1          D2                          D4          D5
D6          D7                          D9          D10
```

Figure 5.15 Workstation display: protection display

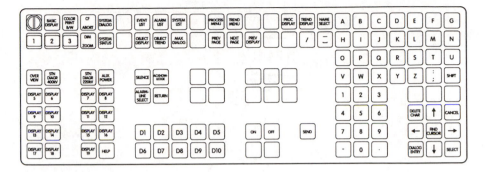

Figure 5.16 Keyboard type 1

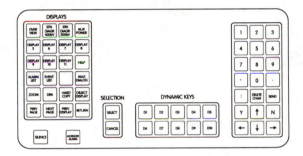

Figure 5.17 Keyboard type 2

5.4.3 Off-line applications

By using a PC, an off-line MMC can be used throughout the system. A portable PC can be connected to any of the REC bay control units for fault tracing and reprogramming. Additional PCs can be used at substation level for calculation and statistical purposes. Relay engineers can use PCs to connect either directly to the SMS in the substation or remotely, in order to display disturbance recording data or examine relay parameters. Portable PCs can also be used by service engineers to test the protection or to simulate disturbances. Figure 5.18 shows a summary of the four possible MMC locations for different personnel: the technician on the protection terminal; the service engineer's portable relay testing and simulation (RTS) system; the operator's on-line workplace; and the relay engineer's on-request workplace.

Since most information is available in digital format this information can be transmitted between each location via an existing telephone system or via diskettes. These off-line functions may also include expert systems for automated data evaluation. Disturbance recording data could, for example, be sent back to the station to test a protection relay with a new setting to be used instead of the old setting that caused maloperation for the same disturbance. This change of setting can also be effected remotely via the SCS or SMS.

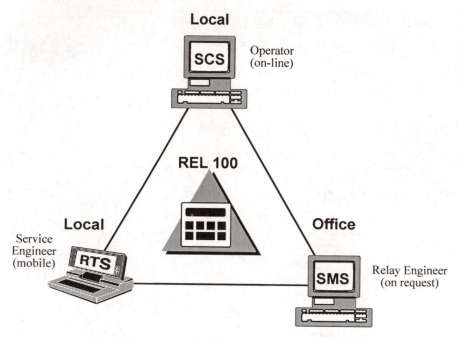

Figure 5.18 Man-machine communication

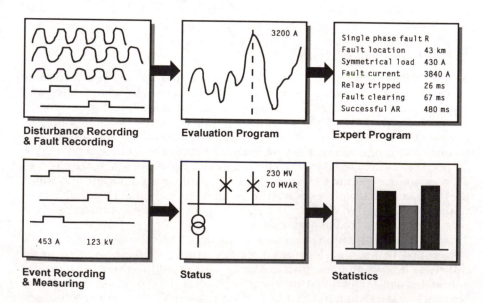

Figure 5.19 Processed information

Figure 5.19 shows a similar procedure for the operator and the protection engineer when "raw" data are evaluated manually, but also used for statistical reports, disturbance analysis or other more "intelligent" functions.

As a summary it can be seen that numerical technology allows for a more "intelligent" operation of the power system where information can be exchanged and processed.

5.5 Advantages of coordinated control systems

The advantages of coordinated control systems can be divided into *short term* during engineering and installation and *long term* during operation and maintenance. The short term is easier to quantify in financial terms compared to the long term which should be a *life cycle cost* comparison.

Before entering into more detailed comparisons it is useful to consider more general issues. Currently, the use of distributed intelligence with microprocessors communicating via fibre optics and the use of PCs as man-machine interfaces is commonplace. Thus new developments are continually taking place within this area hence reducing costs and increasing performance. Furthermore, there are more trained personnel familiar with this new technology. The heavy, centralised, top-down systems used in the sixties and seventies are no longer appropriate since decentralised systems can more easily adapt to future changes compared with centralised systems. In addition, different generations of equipment made by different manufacturers can be easily incorporated. It is easy to upgrade or extend such a distributed system.

A comparison should be made between a conventional substation and a fully automated substation using numerical technology. It is important to make this comparison between "mature" installations. A first trial installation is always expensive. The benefits for the SCS can be summarised as follows for the short term:

- Lower engineering cost by using standardised (user-adopted) software modules;
- Lower installation cost with less cabling and connection points;
- Lower commissioning cost with pre-tested functional modules;
- Lower hardware cost by reduction in the number of auxiliary relays and control boards, and utilising protection hardware for measurement functions;
- Fewer cubicles requiring less space and smaller control housing.

The long term benefits for SCS can be summarised as follows:

- Lower operation cost with remote access down to the bay level;
- Increased availability with self-supervision giving lower interruption cost and lower fault tracing cost;
- Increased lifetime of equipment with scheduled maintenance from statistical data.

Finally several new functions and the possibility of using expert systems will give *added value* which is difficult to define currently in economic terms. With technological changes such as this comes the question of the cost of *not* changing: both spare parts and personnel who are familiar with the old technology may be difficult to obtain in the future.

It is possible to introduce SCS step-by-step, leaving out certain functions, or even have a redundant conventional system. However, to get the full benefit of the new technology and SCS it is necessary to include all functions in the substation: control, protection, monitoring (disturbance recording and signalling), measuring and metering. With the coordinated and distributed concept this can be done without affecting reliability. We can, as an example, imagine control and event recording in one hand, and protection and disturbance recording in the other hand of the same person. They can both be doing different things at the same time but should be *coordinated* so each hand knows what the other is doing. With the new technology it is possible to design the *intelligent substation* for the *intelligent power system*. However, it is important to stress that this will not replace the human operator even if we introduce more automatic functions, expert systems and artificial intelligence. These are only *tools to help personnel perform their tasks more efficiently*.

Glossary

Aliasing	The ambiguous conversion of a high-frequency analogue signal into a low frequency digital signal. See *anti-aliasing filter*.
American Standard Code for Information Interchange (ASCII)	This is a 7-bit code, where the 8th bit can be used for parity check.
Analogue to digital converter (ADC)	A component used to convert an analogue signal into a digital value within a *data acquisition* system.
Anti-aliasing filter	A filter used at the input to a *data acquisition* system in order to ensure that no frequency components greater than half the *sampling frequency* are present otherwise *aliasing* will occur. See *sampling theorem*.
Assembly language programming	Method of programming a microprocessor where individual microprocessor instructions are symbolically represented in a simple programming language. Requires a program called an *assembler* to convert the symbols into machine-readable form. See *high level language programming*.
Asynchronous communication	With asynchronous communication, a character can appear at any given time. When no character is transmitted, the line is in an idle state. Synchronism between the transmitting and the receiving units is maintained throughout the transmission of the character. Transmission begins with a start bit, followed by a character usually in ASCII format, and ends with a parity bit (if used) and a stop bit.

Balanced transmission	The medium is two wires, using two voltages. One voltage is positive and the other is negative. Alternatively, one of the voltages used may be zero volts. The line can have two states, as the wires exchange voltages with each other. Balanced transmission gives greater tolerance to disturbance than unbalanced transmission.
Baud	A unit of measurement for the modulation rate – often the same as the number of *bits per second* (bps).
Bay	A switchgear is divided into bays. Control and protection equipment are often structured the same way. A bay includes all high voltage apparatus such as circuit breakers, disconnectors and instrument transformers. For a $1\frac{1}{2}$ breaker switchgear the bay includes all three breakers and the two connected objects.
Bay level	The name given to a hierarchical level that includes functions which are distributed per bay.
Bit	A binary digit, may assume the values 0 or 1.
Bits per second	Communication rate. See *Baud*.
Bit synchronisation	Used with asynchronous communication and meaning that the clock of the receiving unit is restarted every time a bit is detected.
Boolean algebra	A branch of algebra named after the mathematician George Boole, used to describe the behaviour of logic circuits. See *Karnaugh map*.
Broadcast	The transmission, by a unit, of a message sent simultaneously to several receiving units.
Carrier Sense Multiple Collision Detection (CSMA/CD)	This is a principle used for controlling bus access which is based on collision detection. It is used in the LAN-standard, IEEE 802.3.
CCITT	Comité Consultatif Internationale Télégraphique et Téléphonique. This is an international standardisation body working in the area of telecommunication, which determines the V- and X-standards for data communication. (See below under *V* and *X* series.)
Checksum	A code which accompanies a message. The code is compared at the receiving end to the code at the transmission end. If the code comparison shows agreement at both ends, then it can be assumed that the message has not been altered during the transmission. Checksum can be done with a CRC polynomial.

Combinational circuit	A circuit composed of logic gates that does not possess memory action, and thus has an output which is determined by the combination of its inputs. See *sequential circuit*.
Complementary metal oxide semiconductor (CMOS)	A family of logic devices based upon the switching of a complementary pair of metal oxide semiconducting transistors.
Continuous signal	A signal having the property that the amplitude may continuously change with time. Such a signal cannot be directly processed by microprocessor. See *discrete signal*.
Cyclic redundancy check (CRC)	See *checksum*.
CRC polynomial	A polynomial which is used for checksum calculations. Standardised by CCITT.
Data acquisition	Generic title to describe the process of capturing real-world data into a microprocessor system. Data may be digital or analogue in nature.
Data rate	The average number of bits, characters or blocks which can be transmitted per time unit.
Degradation withstand capability	A measure of the ability of a unit or system to fulfil its function in the event of a partial function fault.
Dependability	The probability that a function will be executed correctly when required.
Digital filter	A filter which acts on sampled values of data. A digital filter is simply a mathematical expression. Two main types: *finite impulse response* and *infinite impulse response*.
Digital signal processor	A type of microprocessor specifically designed for digital signal processing. Usually incorporates an on-board hardware multiplier.
Discrete Fourier transform (DFT)	A mathematical technique for converting a series of time-sampled values into a discrete frequency spectrum. Based on the Fourier transform. See also *fast Fourier transform*.
Discrete signal	A signal having the properties: (1) signal amplitude may only assume one of a sequence of defined levels (discrete level signal); (2) amplitude only defined at discrete points in time (discrete time signal). The term 'discrete time signal' often refers to both discrete time and discrete amplitude representation. Analogue signals must be converted to discrete time form before being processed by microprocessor. See *continuous signal*.

Distributed system

A system in which several individual units in different locations handle system applications. The units are networked together so that they may cooperate.

Duplex

With half duplex, transmission in both directions is possible, but only in one direction at any time. With full duplex, transmission can be in both directions simultaneously. Full duplex transmission requires two separate channels or transmission media.

Dynamic function key

A key (on a keyboard) having a defined function which is shown on the display unit.

Electrically erasable and programmable read only memory (E^2PROM)

A type of ROM where memory contents can be both erased and programmed electrically. Allows change of individual locations.

Erasable programmable read only memory (EPROM)

A type of ROM which can be electrically programmed many times. Memory device must be exposed to ultra-violet radiation for contents to be erased.

Error burst

A number of sequential bits, part of a message, which have been incorrectly transmitted.

Error checking

A method to be certain that incorrect messages are not accepted and interpreted as being correct.

Fast Fourier transform (FFT)

An algorithmically efficient implementation of the *discrete Fourier transform.*

Finite impulse response (FIR)

A type of *digital filter* realised from a non-recursive expression which implies that the filter has a finite impulse response.

Fixed point arithmetic

A method of arithmetic where the decimal point, or its binary equivalent, remains in a fixed position in relation to the microprocessor's data register. Dynamic range of number representation is limited by word length of the data register. See *floating point arithmetic.*

Flash memory

A type of memory that retains its data even if powered down. Blocks of memory can be erased and programmed electrically. Similar, in operation, to *E^2PROM.*

Flip-flop

A component logic device with simple memory action.

Floating point arithmetic

A method of arithmetic where the decimal point, or its binary equivalent, can move, or float, in relation to its position in the microprocessor's data register.

Numbers are thus represented by a mantissa, M, and an exponent, E, and have the general form $M \times 10^E$. Dynamic range of number representation is limited by the number of bits assigned to the exponent. See *fixed point arithmetic*.

Function keyboard A keyboard with function-oriented keys. One key is used to control the same function in several bays.

Graded-index optical fibre A type of optical fibre where the core is constructed from a number of layers of silica each having different refractive indices. This leads to a higher bandwidth than *step-index optical fibres*. See also *single mode optical fibre*.

Hamming-distance This defines the difference between two code words, specifically the number of positions that differ in the code words. A protocol that gives a Hamming-distance of 4 can detect 1, 2 and 3 bit errors, with 100% accuracy.

Hard copy Paper copy.

Hexadecimal A notation for expressing binary numbers by the use of the number base 16. Uses the following characters: 0, 1, 2, 3, 4, 5, 6, 7, 8, 9, A, B, C, D, E and F. Each character represents 4 *bits*.

High-level data link control (HDLC) A data link protocol standardised by ISO.

High level language programming Method of programming a microprocessor where single statements in the programming language allow complicated sequences of processor instructions to be generated. High level languages include C and Fortran. Requires a program called a *compiler* to convert the statements into machine-readable form. See *assembly language programming*.

Infinite impulse response (IIR) A type of *digital filter* realised from a recursive expression which implies that the filter has an infinite impulse response.

The Institute of Electrical and Electronic Engineers (IEEE) An engineering body that assists in developing standards and recommendations.

Interface An interface specifies the connection of a communication link, with regard to the mechanical connection as well as to the electrical and logical characteristics of the signal. An example of an interface is V.24; this has the American equivalent of RS-232.

The International Standards Organization (ISO)	The federation of national standardisation bodies.
Karnaugh map	A pictorial representation of the function of a logic device.
Local area network (LAN)	A local communications network e.g. within a building.
Man-machine communication (MMC)	This refers to the dialogue between the operator and the system; it is used to describe the equipment necessary for this dialogue, such as a video display unit, keyboard, printer etc.
Manufacturing automation protocol (MAP)	This is a combination of several standards and protocols, designated for communication within the manufacturing industry. It is structured according to the OSI model for open systems. See *OSI*.
Modem	Modulator/demodulator. This converts digital data in series format to analogue data in the form of a modulated carrier wave. This is used to extend the distance over which transmission is possible, improve disturbance tolerance and to give galvanic isolation.
Mouse	A hand-held device which is moved across a surface, to indicate various positions on a display unit (SS 011601).
Multidrop	A communication network configured as a bus, i.e. one line that joins together several connecting junctions.
Multiplexer (MUX)	A device which allows several signal inputs to share a single output in a sequential fashion. Uses include (1) *data acquisition* systems where several analogue signals are multiplexed into the same *analogue to digital converter,* and (2) communication networks where several low bandwidth signals are shared on a high bandwidth medium.
Object bus (OB)	This is a communication link between the station bus interface and functional units/ multifunctional units at the bay level.
Object bus interface	Functional units/multifunctional units are connected to the object bus via the object bus interface.
Operating location	The working location of an operator, equipped with a video display unit, location keyboard etc., for controlling and monitoring a process.

Open Systems Interconnection (OSI)	A model for communication protocols, divided into 7 layers, each having a specific purpose. The idea is to make easier the communication between various manufacturers' equipment, when the interface to each layer is standardised. Layer 1: physical; layer 2: data link; layer 3: network; layer 4: transport; layer 5: session; layer 6: presentation; layer 7: application.
Parity	Parity check is used at the character level of a message. This means that a parity bit is transmitted together with a group of bits, and is 1 if the total number of 1s in the group is even, and zero otherwise (even parity). The opposite is true when odd parity is used.
Poll	A master "questions" a slave and waits a specified time to get an answer.
Polled system	A system whereby a master unit cyclically checks a number of slaves.
Programmable read only memory (PROM)	A type of *ROM* which can be electrically programmed once.
Protocol	A set of rules regarding the exchange of information.
Random access memory (RAM)	Memory used for transient data store — contents can be changed. RAM contents lost if powered down.
Read only memory (ROM)	Memory used for permanent data store – contents cannot be changed. Data retained even if powered down.
Real time	Year, month, day and time, according to Greenwich Mean Time (GMT) or local time.
Relative time	A time relative to a synchronising pulse or another event.
Remote terminal unit (RTU)	Used with regard to remote control terminal. It is generally an unintelligent in/out unit that is connected to the system via the communication link to a computer system.
Resolution	The smallest representable part of a measurement.
Response time	The time between when a command is given by the operator and a response, indicating execution of the command at the bay level. It can also be the time between an automatic command to a unit and the time when the receiving unit has executed the command.

RS-232-C	An Electronics Industries Association (EIA), USA standard. It corresponds to CCITT V.24, V.28 and ISO 2110 (mechanical standard) in combination.
RS-422	A standard for balanced transmission having a high data rate.
RS-423	A standard for unbalanced transmission for high data rates.
RS-485	A standard for electrical interfaces for balanced transmission, in a multi-drop network. This is a further development of RS-422. The maximum data rate is 10 Mbps. The maximum length is in the kilometer range. 32 units can be joined to the network. The allowed voltage levels of -7 V to +12 V mean that a supply voltage of 5 V can be used. See *balanced transmission*.
Relay testing and simulation system (RTS)	RTS is a computerised system for testing and simulation of the protection functions. Automatic testing of, for example, a line protection or replay of a disturbance from a digital disturbance recorder is possible. It is also possible to edit the waveforms before testing.
Sample and hold amplifier (SH)	A component within a *data acquisition* system used to 'hold' the input analogue signal at a constant level whilst the *analogue to digital converter* performs its task.
Sampling	The process of converting analogue signals to a digital representation. Digital values referred to as samples. See *data acquisition, analogue to digital converter*.
Sampling frequency	The rate at which *sampling* is performed.
Sampling theorem	Theorem that dictates that *sampling* must occur at, at least, twice the highest signal frequency.
Sequential circuit	A circuit composed of logic gates that possesses a memory action and thus has an output which is determined by a sequence. See *combinational circuit*.
Single mode optical fibre	A type of optical fibre having a small cross-sectional area. This ensures propagation of light rays, with large angles of reflection, in one mode only. See *graded-index optical fibre, step-index optical fibre*.
Step-index optical fibre	A type of optical fibre where the core and cladding have different refractive indices. Propagation is thus due to total internal reflection at the interface. See *graded-index optical fibre, single mode optical fibre*.

Substation monitoring system (SMS)	SMS includes various monitoring functions in a substation. It is an off-line system which is not time-critical, unlike SCS. It does not include any control functions except changes of parameters for the protective relays. The main functions in SMS are disturbance recording and other functions required by the relay engineer. SMS can be interconnected or can be a subset of SCS.
Substation control system (SCS)	SCS includes all control and supervisory functions in a substation as well as interlocking of circuit breakers and disconnectors. SCS can replace both the conventional RTU and the switchboard in the substation. Many additional features such as event-logging, alarm-handling, statistics and sequential control functions such as automatic restoration can be included.
Synchronous data link control (SDLC)	A protocol developed by IBM for a master/slave system. This is built upon a part of HDLC.
Security	This is the probability that a function will not be performed when unwanted.
Simplex	Data transmission whereby transmission is possible in only one direction at any one time.
Station bus (SB)	A communication link between the station level and the bay level.
Station bus interface (SBI)	The bay's interface with the station bus.
Synchronous communication	The transmission of data from one place to another under regular fixed time intervals.
Terminal	A functional unit which can communicate with the substation control system and/or substation monitoring system. An object terminal refers to a specific object, e.g. transformer terminal, line terminal and generator terminal. A terminal can include any or all of the following main functions: protection, control, measuring and monitoring, as well as additional related functions. A line protection terminal may, as an example, include distance protection, earth fault protection, fault location, load measurement and disturbance recording, all of which can communicate with the SCS and/or SMS.
Token bus	Access to the bus is controlled by means of a "token" which is passed around the units connected to the network. A unit having the token is permitted to transmit data, while all other units wait for their turn. This principle is standardised in IEEE 802.4, which is the basis for MAP.

Token ring	A ring-formed network, having the same principle as the token bus. Standardised IEEE 802.5.
Transistor-transistor logic (TTL)	A family of logic devices based upon the switching of bipolar transistors.
Turnaround time	The time for a master to make one cycle, polling all slaves.
Twos complement	A method of assigning values to binary numbers that allows both positive and negative numbers to be represented by taking the most significant *bit* to be negative.
V-series	CCITT's recommendations for data transmission via a telephone network.
V.10	Conforms to RS-423, unbalanced transmission.
V.11	Conforms to RS-422, balanced transmission.
V.24	Standard for the connection of computers/terminals to a modem. Used together with V.28 electrical interface.
V.28	A standardised electrical interface for the connection of computers/terminals to a modem. Signal levels of ±12 V, transmission data rate maximum is 19 200 bps, 25-terminal connecting unit, maximum distance is 15 m.
X-series	CCITT's recommendations for data transmission via a digital network (a public network).
X.21	Standardised electrical interface for connection of computers/terminals to a synchronous network.
X.25	A protocol used for communication between computers/terminals on packet-switched networks.

Index